Island Geographies

Islands and their environs – aerial, terrestrial, aquatic – may be understood as intensifiers, their particular and distinctive geographies enabling concentrated study of many kinds of challenges and opportunities. This edited collection brings together several emerging and established academics with expertise in island studies, as well as interest in geopolitics, governance, adaptive capacity, justice, equity, self-determination, environmental care and protection, and land management. Individually and together, their perspectives provide theoretically useful, empirically grounded evidence of the contributions human geographers can make to knowledge and understanding of island places and the place of islands. Nine chapters engage with the themes, issues, and ideas that characterise the borderlands between island studies and human geography and allied fields, and are contributed by authors for whom matters of place, space, environment, and scale are key, and for whom islands hold an abiding fascination. The penultimate chapter is rather more experimental – a conversation among these authors and the editor – while the last chapter offers timely reflections upon island geographies' past and future, penned by the first named professor of island geography, Stephen Royle.

Elaine Stratford works at the University of Tasmania, Australia, where she is Professor and Director of the Peter Underwood Centre for Educational Attainment, and is affiliated with the Discipline of Geography and Spatial Sciences in the School of Land and Food.

Routledge Studies in Human Geography

This series provides a forum for innovative, vibrant, and critical debate within Human Geography. Titles will reflect the wealth of research which is taking place in this diverse and ever-expanding field. Contributions will be drawn from the main sub-disciplines and from innovative areas of work which have no particular sub-disciplinary allegiances.

For a full list of titles in this series, please visit www.routledge.com/series/SE0514

Everyday Globalization
A spatial semiotics of immigrant neighborhoods in Brooklyn and Paris
Timothy Shortell

Releasing the Commons
Rethinking the futures of the commons
Edited by Ash Amin and Philip Howell

The Geography of Names
Indigenous to post-foundational
Gwilym Lucas Eades

Migration Borders Freedom
Harald Bauder

Communications/Media/ Geographies
Paul C. Adams, Julie Cupples, Kevin Glynn, André Jansson and Shaun Moores

Public Urban Space, Gender and Segregation
Women-only urban parks in Iran
Reza Arjmand

Island Geographies
Essays and conversations
Edited by Elaine Stratford

Island Geographies

Essays and conversations

Edited by Elaine Stratford

Routledge
Taylor & Francis Group

LONDON AND NEW YORK

First published 2017 by Routledge

2 Park Square, Milton Park, Abingdon, Oxfordshire OX14 4RN

711 Third Avenue, New York, NY 10017

Routledge is an imprint of the Taylor & Francis Group, an informa business

First issued in paperback 2018

British Library Cataloguing in Publication Data
A catalogue record for this book is available from the British Library

Library of Congress Cataloguing in Publication Data
Names: Stratford, Elaine, editor.
Title: Island geographies : essays and conversations / edited by Elaine Stratford.
Description: New York, NY : Routledge, 2017. |
Series: Routledge studies in human geography |
Includes bibliographical references and index.
Identifiers: LCCN 2016023775| ISBN 9781138921726 (hardback) |
ISBN 9781315686202 (ebook)
Subjects: LCSH: Islands–Case studies. | Human geography–Case studies.
Classification: LCC G471 .I75 2017 | DDC 910.914/2–dc23
LC record available at https://lccn.loc.gov/2016023775

ISBN: 978-1-138-92172-6 (hbk)
ISBN: 978-1-138-33935-4 (pbk)

Typeset in Times New Roman
by Out of House Publishing

To the memory of Kate Stratford [née Whalley] (1925–97) and Will Stratford (1926–99), from whom the editor first learned both the value of higher education and the warmth emanating from collaboration.

'Islandness is an intervening variable that ... contours and conditions physical and social events in distinct, and distinctly relevant, ways.' (Baldacchino, 2004, p. 278)

'... there are many reasons why islands are and should be of growing interest to geographers.' (Mountz, 2014, p. 8)

Contents

List of illustrations ix
Notes on contributors x
Acknowledgements xiii

1 **Introduction** 1
ELAINE STRATFORD

2 **The deep Pacific: island governance and seabed
mineral development** 10
KATHERINE GENEVIEVE SAMMLER

3 **Islands and lighthouses: a phenomenological geography of Cape
Bruny, Tasmania** 32
THÉRÈSE MURRAY

4 **Too much sail for a small craft? Donor requirements, scale, and
capacity discourses in Kiribati** 54
ANNIKA DEAN, DONNA GREEN, AND PATRICK D. NUNN

5 **An island feminism: convivial economics and the women's
cooperatives of Lesvos** 78
MARINA KARIDES

6 **Nature and islands: rethinking the cultural heritage of
New Zealand's protected islands** 97
DAVID BADE

7 **'The good garbage': waste-to-energy applications and issues
in the insular Caribbean** 114
RUSSELL FIELDING

8   **The returning terms of a small island culture: mimicry, inventiveness, suspension**   132
    JONATHAN PUGH

9   **Conversations on human geography and island studies**   144
    ELAINE STRATFORD AND AUTHORS

10 **Retrospect and prospect**   160
    STEPHEN ROYLE

    Bibliography   169
    Index   192

Illustrations

Figures

2.1	Select Pacific Island nations: land and territorial areas	18
2.2	Territorial contiguity in the Pacific created by EEZ jurisdictional borders	21
2.3	Proposed mining sites, New Zealand's EEZ and ECS marine territory	24
3.1	Cape Bruny Lighthouse	34
3.2	Skeleton	42
3.3	A bridge across space and time	44
4.1	The Republic of Kiribati	57
4.2	Erosion on the coastline adjacent to the Taborio-Ambo seawall, Kiribati	71
6.1	New Zealand: location of select islands	99
6.2	Rangitoto as seen from Auckland, New Zealand	107
6.3	The first volunteer-planted forest of Motutapu, located at the Home Bay Valley	112
7.1	Locations of St Croix, St Barthélemy and surrounding islands	116
7.2	A section of the 'Jepp chart' showing the Henry E. Rohlsen Airport in St Croix	123

Tables

4.1	Physical investments from KAP II (2006–11)	71
5.1	Lesvos women's cooperatives: village, year initiated, and membership	87

Contributors

David Bade completed his PhD in human geography at the University of Auckland in 2013. His thesis was entitled 'Managing Cultural Heritage in "Natural" Protected Areas: Case studies from Rangitoto and Motutapu islands in Auckland's Hauraki Gulf'. He has worked as a Specialist – Built Heritage in the Built & Cultural Heritage Policy Team at the Auckland Council, New Zealand, and as Government Advice Project Officer (Research & Planning) at Historic England.

Annika Dean is interested in issues of community development, disaster risk reduction, and climate change adaptation in the Pacific region, particularly in terms of how aid and climate finance impact adaptive capacity and affect climate change adaptation initiatives in Pacific island countries. Annika is affiliated with the Climate Change Research Centre (CCRC) of the University of New South Wales, Australia.

Russell Fielding is an Assistant Professor of Environmental Studies at the University of the South in Sewanee, Tennessee, USA. His graduate degrees in geography are from the University of Montana and Louisiana State University. He spent a year as a Fulbright Scholar at the Institute of Island Studies at the University of Prince Edward Island in Canada. Fielding is interested, broadly, in questions of subsistence, cultural tradition, and resource conservation. His current research investigates local methods of food and energy production on small islands of the Caribbean and North Atlantic.

Donna Green is a senior research scientist in the Climate Change Research Centre, University of New South Wales, Australia. In this position she leads a program that explores how indigenous and non-indigenous knowledge can be used to understand climate impacts in remote communities. Her research focuses on human–environment interactions, specifically on social and economic vulnerability, adaptation, and risk.

Marina Karides is Associate Professor of Sociology at the University of Hawaii, Hilo. Her research combines the fields of globalisation, social inequalities, and global social movements with island studies, and thus is inflected with key geographical concerns. She has completed fieldwork in the Republic of Trinidad and Tobago, the Republic of Cyprus, Lesvos, Greece, and Hawaii Island, and published works that examine how alternative, small-scale, and local economies on islands intersect with gender and sexuality and oppose neoliberalism. Her forthcoming book is *Sappho's Legacy: Convivial Economics on a Greek Isle.*

Thérèse Murray studied geography and philosophy at the University of Tasmania, Australia. She is interested in geographies of place and in understanding in how specific places are felt and experienced. Thérèse is also interested in the material effects of geographic imaginaries on human lives and inequity with particular regard to asylum seeking and imaginaries of island and ocean.

Patrick D. Nunn is Professor of Geography at the University of the Sunshine Coast, Australia, and earlier spent 25 years at the international University of the South Pacific where he had opportunities to undertake research in most Pacific Island countries. He has been involved in climate change research in the Pacific for most of his career and has authored more than 230 peer-reviewed publications. He has been a member of the Intergovernmental Panel on Climate Change (IPCC) on several occasions, was a co-recipient of its 2007 Nobel Peace Prize, and a Lead Author of the most recent Assessment Report chapter on 'Sea Level Change'. His current research focuses on community-livelihood sustainability in the Asia-Pacific region.

Jonathan Pugh is Senior Academic Fellow at the School of Geography, Politics and Sociology, Newcastle University, and Honorary Fellow of the Centre for the Study of Democracy, University of Westminster, UK. He is a geographer who focuses upon questions of impasse and suspension in critical theorising of the everyday, island and archipelagic studies, and radical political philosophy and politics.

Stephen Royle studied geography at St John's College, Cambridge University and undertook his PhD at the University of Leicester, UK. He taught geography at Queen's University Belfast in Northern Ireland from 1976 to 2015, retiring as Professor of Island Geography. He wrote the chapter for this book whilst Visiting Professor at the Kagoshima University Center for Pacific Island Studies, which reflects his island travels – some 843 of them visited to date. Royle's books include *A Geography of Islands* (2001), *The Company's Island* (2007), *Company, Crown and Colony* (2011), and *Islands* (2014). He was a founder member and is treasurer of the International Small Island Studies Association, and serves as book review editor for *Island Studies Journal*. Royle was elected as a Member of the Royal Irish Academy in 2008.

Katherine Genevieve Sammler is affiliated with the School of Geography and Development at the University of Arizona, USA. She is a political geographer whose current research focuses on territory, sovereignty, and resource governance in frontier and less-than-terrestrial spaces like the deep ocean and outer space.

Elaine Stratford is Professor and Director of the Peter Underwood Centre for Educational Attainment at the University of Tasmania, Australia, and internationally known for her work in cultural and political geography and island studies, not least with children and young people. Elaine's research is motivated by trying to understand the conditions in which people flourish in place, in their movements, in daily life, and over the life-course. Much of such work is presently directed towards leading efforts to champion educational attainment and raise aspirations for lifelong learning among children and young people in Tasmania.

Acknowledgements

Elaine Stratford wishes to acknowledge her collaborators herein, most of whom first started to work on this project at the 2014 Association of American Geographers Conference held in Tampa, Florida. Individually and collectively, the authors have engaged with Stratford to craft separate essays and shared conversations, and have given to this project more than might usually be expected. Particular thanks to Thérèse Murray for her stellar organisational skills in navigating the contours of international telecommunications networks that enabled us all to speak over vast distances. Thanks, too, to the team at Routledge for inviting Stratford to submit the initial proposal for this edited collection and for being open to the conversational elements that have been central to her desire for an overarching coherence to the work in total.

Katherine Sammler notes that the work in this book is drawn from research supported by the National Science Foundation under Grant No. 1415047. Sammler also benefitted from a Social Science Research Council Fellowship, which engendered rich discussions with colleagues in the 'DPDF program: Oceanic studies: Seas as sites and subjects of interdisciplinary inquiry'. Assistance was also provided by the University of Arizona's Social and Behavioral Sciences Research Institute and the Graduate and Professional Student Council. Special thanks to research participants and Dr Nick Lewis at the University of Auckland, and to Carly Nichols, Megan Mills-Novoa, and Sallie Marston for encouraging and thoughtful comments on the draft of the chapter, made during a graduate seminar on writing for publication.

Thérèse Murray would like to thank her colleagues at the University of Tasmania who generously contributed their thoughts and experiences of place at Cape Bruny, Tasmania. Thanks, too, to Thérèse's parents for a lifetime of support and, in particular, her father who gifted her an enduring fascination with geography.

Annika Dean, Donna Green, and Patrick Nunn would like to acknowledge the Climate Change Research Centre at the University of New South Wales and Professor John Connell from the School of Geosciences at the University of Sydney, Dr Paul Hodge from the School of Environmental and Life Sciences at the University of Newcastle, and Dael Allison, each for providing valuable feedback and edits on a first draft of the chapter. Acknowledgements also go to Claire Anterea, for help and support in the field, and to all of the people who participated in the research.

Marina Karides is grateful for funding assistance from the Fulbright Scholar Program (Greece). Her gratitude goes to Dr Marilyn Brown and Dr Lindy Hern, who were especially supportive of her joining the University of Hawai'i at Hilo, giving her the intellectual space to develop island feminism. The comments of the editor, Elaine Stratford, and other contributors to the volume are much appreciated for improving the manuscript. She feels especially indebted to the cooperative members for the time granted for interviews and permission for observations. Not to go without mention are thanks for the delicious pastries and savories generously offered by all the cooperatives.

David Bade would like to acknowledge his two PhD thesis supervisors, Dr Gretel Boswijk and Professor Robin Kearns of the School of Environment, University of Auckland, New Zealand, his parents James and Margaret, brother Richard, and wife Stephanie for their support.

Russell Fielding acknowledges local informants on St Barth: Hélène Bernier, Alexandra Deffontis, Jean-Philippe Piter, and Christophe Turbé. Special thanks to Président Bruno Magras for taking time to meet with a visiting geographer. María Elena Díaz, of Columbia University, and Don Hurd, of Alpine Energy in Denver, provided important information on waste-to-energy facilities. Adam Dahl at Sewanee shared insights on political theory that strengthened the comparative analysis reported in the chapter. Matt Moseley provided an insider's view on challenges to aviation presented by the Anguilla landfill and helped decipher the Jepp charts. Further grounding discussions came from contributors to the sbhonline.com forum. Thanks to the editors of *Focus on Geography*, for allowing material published in the journal in 2014 to be reprinted here. Special thanks to the Office of Internationalization at the University of Denver and the University Research Grants program at the University of the South, which provided funding for fieldwork.

Jonathan Pugh notes that this chapter grew out of work supported by a Caribbean: Economic and Social Research Council (ESRC) PhD studentship 1998–2001 (R00429834850); ESRC three-year fellowship 2002–4 (R00271204); Research Council United Kingdom fellowship 2005–10 (EP/C509005/1); and Newcastle University School of Geography, Politics and Sociology Research Fund (2011). Thanks to Alison Donnell and Malachi Mcintosh for conversations about Caribbean literature, politics, and geography, and for reading drafts of the chapter. A version of this work was presented as a keynote address at the International Small Island Studies Association, Islands of the World Conference XIII, Penghu Islands (Pescadores), Taiwan Strait, September 2014.

1 Introduction

Elaine Stratford

In 1972, having lived in Saskatoon, Saskatchewan for over a decade, my parents felt called to New Zealand – my father, Will, to undertake his doctorate in education at the University of Otago, my mother, Kate, to establish a tertiary-level speech pathology training course in Christchurch. It was the second time they had been drawn to those islands. The first such occasion, in 1950, involved three £10 passages – for them and my brother William – and a long voyage from their home in the British Isles via the Atlantic Ocean, Caribbean Sea, and Panama Canal, and south through the Pacific Ocean to New Zealand's North Island. There, they made their way to a small hamlet called Mokauiti, 'as the crow flies' about 89 kilometres (56 miles) from the Waikato regional capital of Hamilton. In that remote place, and for just over three years, Will was the sole teacher in a two-roomed school, and Kate minded both William and my sister Caroline, born (literally) at the school in 1952 – so remote the settlement and so muddy the roads that the doctor could not get through in time to deliver her. But in 1953, the family returned to the north of England, later adding sister Nicola to the mix, before realising that life in the UK remained parlous in the reconstruction period following World War II. So they headed off again, this time to the backblocks of rural Saskatchewan, and to the embrace of another hamlet, Porcupine Plain. The changing needs of pubescent children, and more varied opportunities to practise teaching and speech pathology then enticed them to Saskatoon, 280 kilometres (173 miles) away where, in time, I arrived.

Growing up, I listened avidly to conversations about those adventures: my parents' very different homeland villages nestled in the beech forests of Buckinghamshire and the moors of Yorkshire; their meeting in London after the war; and the compulsion to seek better lives elsewhere; the delights and mishaps of ocean travel; life in the forested hills of New Zealand; and the sense of being happily enisled on the prairies – an observation I heard again from Minnesotan islophile Bill Holm (2002), for whom prairie settlements were islands in grainy seas. And then, in 1972, the impulse to move returned, the decision made – New Zealand, the sequel. But with the two older children – now young adults – opting to stay in North America, just four made the trip south in two stages. First, my father, sister Nicola, and I left autumnal

Saskatoon in the August of 1972, destined for Dunedin and a Canada Council PhD scholarship for him, but taking over a month to get there via four Pacific islands. Then my mother joined us nine months later having secured the afore-mentioned post in Christchurch.

Thus it was that in August 1972 I found myself standing at some point on the coast in Vancouver, where views of the Pacific are sweeping, and I looked up at my father asking where this vast ocean ended. Where we are going to live, he said, at the other end of the Earth ... and he teased me about learning to walk upside down. And I recall days and days of smiling islanders, warm sun, 'mocktails' with maraschino cherries speared on kitsch paper umbrellas, palm trees and swimming pools, coconuts and yams, the novel taste of fresh seafood, the roughness of coral and grittiness of sand, and a sense of delight at being able to walk around an island in one go and take in what I thought was its entirety. The seashore, that liminal space between, then fascinated me; it still does. The odd mix of fantasy and verisimilitude between what I could see with my own eyes and what I could see in paintings by Gaugin in the museum in Tahiti or old maps on the walls of a small museums in Samoa ... this, too, captivated me.

Formative though these times were, their significance was made more powerful by one experience had during 12 days in Nandi, the importance of which was to take years to register. By the time we arrived in Fiji, my sense of wonderment had been tempered by a visceral understanding that my mother was a long way from me and would be so for some time; that older siblings, too, were 'gone'; and that I was going somewhere 'foreign' to live, and would I like it? Little wonder that I took to the Fijian maid, Mary, who was about my mother's age and who quietly tolerated my following her around the bunga-lows at our little resort, one hand of mine softly gripping her crisp uniform or white pinafore. Not really apprehending the idea of a 'rostered day off', I felt forlorn when, for two days in a row, Mary did not appear, but gladdened when sister Nic, to me wise at 14, suggested we walk around the district and see if we could find her. And so we went rambling and, indeed, we found Mary on the outskirts of the settlement, in an area of ramshackle buildings made of corrugated iron and chip- and cardboard. Mud puddles pockmarked the pathways, small children in scant clothing played in dark doorways, and my new patent leather shoes and floral dress seemed entirely out-of-place. I now have a clear recollection of feeling deep unease – later it would register as the question *how could I live the way I did and they in this way?* It felt deeply wrong, although at that time I could not articulate why. Later, in uncompli-cated fashion, I thought about 'haves' and 'have-nots'. Later still, I learned to question the ethical basis that initially compelled me to give superordinate value to my life and its privileges; then recognise the need to consider certain cultural and spatial *relativities*; and then acknowledge the *absolute* injustice of the deprivations I witnessed that day.

In retrospect, I think my lifelong engagement with geography began during that trip south from Canada to New Zealand, but it took years to emerge as a conscious and political focus; years during which another move was made to the island-continent, and to Adelaide, where I completed my schooling and enrolled in courses at both Flinders and Adelaide Universities – majoring in geography and visual arts, teaching, and taking a PhD in environmental studies. And then, in 1996, I made one final move with my own growing family to my first (and current) substantive academic position at the University of Tasmania, in the island state of the island-continent; my home for two decades now. It was here that my commitment to human geography – and especially to questions of how we flourish over the lifecourse – finally connected with my fascination with islands, which in truth and to that point had been the stuff of childhood memories.

From early 1997 to the present time, I have increasingly focused upon island studies and its relationship to human geography, and my work has come to be informed by a desire to consider the challenges faced by island peoples. Some of these – such as the relative poverty in which Mary and her family lived – are not unique to islands, but perhaps they seem pronounced in small land- and seascapes. Some of them seem particular to islands – though I am never fully persuaded I know with unshakeable certainty which ones. Perhaps for me the most obvious of these challenges has been the effect of sea-level rise on areas of land so small that centimetres rather than metres of change have significant effects – not least for island children in the future (Stratford *et al.*, 2013; Stratford and Low, 2015).

During the intervening years, I have been fortunate to attend many of the biennial conferences of the International Small Island Studies Association, and to be involved in organising or participating in island sessions at geography conferences in the United States, Australia, and the United Kingdom. From such collaborations have come many insights and just as many questions, and these have gained expression in, for example, a journal special issue (Stratford, 2013), a book series entitled 'Rethinking the Island' (Rowman and Littlefield International), and this present collection of essays and conversations.

Most recently, undertaking a ten-year review of the content and impact of the *Island Studies Journal*, I discerned that the field is characterised by high levels of interdisciplinarity, which is evinced by significant input to the journal from those based in economics and management, earth and environmental sciences, and anthropology, and archaeology, among others. Yet those 'authors primarily identifying as geographers, or affiliated with, geography departments in higher education organizations are most numerous' in the journal (Stratford, 2015, p. 145). At the same time, the preponderance of geographers writing about islands is, I suggest, quite straightforward: as Royle notes later in Chapter 10, geographers have been, and many remain, primarily concerned with exploring place, space, and environment. Explicitly many of us are also fascinated with questions of scale, the impact of movement and stasis and states in-between, and with the particularities of the *human* in geographical materialities and imaginaries. The island, the archipelago, islandness, the effects and affects of land and of water, and of movements vast, large, and small – such matters are both profoundly geographical, and central to the essays that follow.

This edited collection began as a series of papers in three sessions at the 2014 conference of the Association of American Geographers held in Tampa, Florida. It results from the generosity and disciplined focus of each of the contributing authors – those present at the Tampa meeting and those who joined us after the fact and helped round out the offerings. As editor, I have been particularly keen to capture the reality that our conversations with each other have extended beyond the boundaries of the conference venue, and include many emails to-and-fro about the original proposal, drafts of chapters, and our shifting thinking about our ways of 'doing geographies' and understanding islands over the ensuing period.

Taken individually these essays have, I think, useful contributions to make to the scholarship of island geographies. Collectively they are, I hope, greater than the parts. If one were to read through them at a sitting, as I have done (and on more than one occasion, of course) obvious and prevailing themes and issues emerge. Across the chapters some of them are geographical, and pertain to space, place, and environment – as Royle revisits later in the work. Some relate to scale and to questions about what happens in, between, and across territory, micro-site, site, locale, region, sub-national jurisdiction, or nation-state. Some themes and issues concern certain mobilities, frictions, proximities, distances, and flows – not least in the exercise of powers that are individual, collective, commercial, or sovereign, juridical, informal, or tacit. Others relate to questions of identity, indigeneity, settler status, gender, race, class, socio-economic status, political bent.

And as I read and reread the lists I have written above, and as I think about their function, which is to labour for indexical and categorical purposes, I am struck again by Baldacchino's assertion, invoked and tested in several of his works, that islands are not simply intervening variables but amplifiers and intensifiers also. Therefore, I discern that more nuanced ideas also emerge from engaging with the essays, which variously consider how island geographies tend to create different networks, assemblages, relations, gatherings, bridges and, in addition, diverse ruptures, boundaries, gaps, and absences. These ideas are to be found in each of the works in varied ways as the authors productively struggle with some of the historically (and confounding) key concepts in island studies: size, isolation, remoteness, boundedness, insularity, openness, access, identity, marginality, dependence, self-sufficiency, and resourcefulness.

Nine chapters follow, and in different ways and to varied extents, the authors engage with the themes, issues, and ideas that I have outlined above. Of these chapters, seven are essays by established and emerging scholars in human geography and allied fields for whom matters of place, space, environment, and scale are key, and for whom islands hold an abiding fascination. One chapter is rather more experimental – a conversation – and one is a perspectival offering that reflects upon island geographies' past and future, penned by the first named professor of island geography.

As might be expected by readers, the order of appearance of chapters is not arbitrary, and involved my having thought about key geographical concepts such as space, place, and environment as well as others including scale, locale, and region. For me, it was important to start with the seas. Emphasising the ocean as elemental, central, delineating in studies of island geographies is important, and certainly indebted to insights provided by Hay (2013) and Hayward (2012), at least in part in response to work I have done with colleagues on how to think about the island and the archipelago (Stratford *et al.*, 2013). The first writes: 'if there is enough substance to the notion of islandness to justify a coherent intellectual preoccupation called "island studies", it must have to do with water, the element common to all islands, and, more specifically, the sea' (Hay, 2013, p. 211). Consider, then, the suggestion that an intelligible view of islandness requires critical acceptance that there are certain island psychologies that integrate containment, remoteness, and isolation, and that 'must have to do with the element of the sea [as a] primary condition' of islandness (Hay, 2013, p. 209).

The second works to explain his call for a neologism, *aquapelago*, on the basis that 'it is possible to argue that the word "archipelago" islands as land masses to be useful as a designation for is now too heavily associated with concepts of regions in which aquatic spaces play a vital constitutive role' (Hayward, 2012, p. 5). Thus consider, too, his description of island studies as:

> research into island communities – social entities that have both an insular condition, being surrounded by sea, and, usually, a connectivity, produced by the use of the sea as a means of navigating between islands and/ or mainlands. Island communities … are innately linked to and dependent on finite terrestrial resources and constantly react to and work within the transitional zone between land and sea, in the form of the shoreline and adjacent coastal waters and more distant and deeper marine environments. These are the defining aspect of their geographical and geo-social identities.
>
> (Hayward, 2012, p. 1)

Hence in Chapter 2, Katherine Sammler explains how economic motivations and technological advancements have made possible deep seabed mining and the removal of precious metals from the sea floor, which may increase toxicity and turbidity in the water column. Sammler speculates that Pacific island nations may gain the most from developing their spatially significant Exclusive Economic Zones (EEZ). Among such nations, New Zealand recently approved steps for mining its seabed ironsands, and will allow its Defence Force to arrest and detain anti-mining protesters in its EEZ. Stakes are high for state and corporate interests, and for those arguing the legitimacy of these resources as a public asset. Institutional nascence and untested mining technologies come together to produce a dangerous socio-ecological experiment. Drawing on insights from political geography, the chapter considers the question how are conventional environmental governance schemes being reconfigured to confront ocean resource governance, and to what extent are issues of economic fairness and environmental impacts being addressed?

Sammler's concerns are with deep oceans, their resources, and their effects on island and archipelagic nations. In turn, in Chapter 3 Thérèse Murray has as her immediate focus an iconic coastal structure: the lighthouse at Cape Bruny, Tasmania. Murray's analysis – embracing land, water, weather, and lives – brings to mind work by Hester Blum (2013, p.151), who has argued that 'the sea should become central to critical conversations about global movements, relations, and histories'. Surely the material and symbolic influences of the lighthouse could be elemental in the conversations Blum invites – for, as Murray demonstrates, along with its grounds, keepers, and others this structure exemplifies the relationships of people to place, of ships to shores, and of islands to imperial expansion across the globe. Taking an aspect of Martin Heidegger's thought regarding the relationship between building and the gathering of the world into places as a launching point, Murray seeks to gain insight into whether and how the essential nature of built structures becomes entwined with places (and, indeed, with the mobilisation of place). Drawing broadly on phenomenological approaches to qualitative research in human geography, the work tracks various individuals' perceptions, experiences, and insights, and simultaneously explores the lighthouse and the place in which it stands, advancing a series of geographical reflections on the profoundly important relationships between bodies of water, islands, and lighthouses.

In Chapter 4, other questions about bodies of water, infrastructure projects, island peoples, and continental aid agencies concern Annika Dean, Donna Green, and Patrick Nunn. There, with specific reference to the Pacific island nation of Kiribati – an archipelago straddling the Equator – they examine islands' vulnerability to the impacts of climate change, and consider the ways in which the administration of climate finance can lead to compound injustices. They acknowledge that the peoples of small island developing states and – specifically – Pacific island populations have done little to cause climate change but will face significant, sometimes crippling, adaptation and mitigation costs. In response to such inequity, governments in developed countries have agreed to mobilise climate finance to enable adaptation and mitigation. Yet Pacific island governments have faced difficulties accessing such finance or, and in the case that Dean and her colleagues discuss, have faced challenges dealing with the allocation of funds in the terms provided by donors. Addressing such issues, Dean and her colleagues draw on empirical data collected during fieldwork in Kiribati. Their work considers a three-stage project funded by the World Bank and finds that perceived lack of capacity among Kiribati's recipient populations is a key challenge. The authors posit that, by holding a tight rein over project-based external finance, donors run certain risks: draining capacity; disempowering and demoralising the I-Kiribati; and entrenching a postcolonial paternalism that ironically could undermine resilience to climate change. Such modes of project delivery and their impacts seem to compound and add to the original injustices of climate change, outcomes that the authors suggest now warrant a radical rethinking of climate finance in island places.

Marina Karides considers another situation in which radical changes are likely required. In Chapter 5, Karides develops a framework for island feminism and argues that richer critical understandings of islands depends on appreciating and assessing the gendered experiences and organisation of island communities. For Karides, island feminism refers to the intellectual sensibilities of island place and constructs of gender and sexuality, and she views them as interlacing forces that shape economic, social, and ecological life, and cultural and political conditions that apply to islands. Island feminism reappropriates narrow understandings of 'island women', and is informed by the dynamic interplay of culture, place and space, and identity as these manifest in what Karides calls convivial economic practices. Responding to what she sees as significant gaps in island feminist scholarship, and drawing on field work and interviews conducted on Lesvos, Greece, between 2008 and 2012, Karides offer a case study of 11 of the women's cooperatives on the island. Her aim is to constitute what she calls a nissological feminist geography to highlight island economic strategies that have secured, and could continue to bolster, financial well-being and provide for the survival of island communities.

The vulnerability of other peoples, and of particular forms of cultural heritage in natural protected areas, is the focus of Chapter 6. Writing about these matters by reference to New Zealand's offshore islands, David Bade describes how, in the 1890s, certain of New Zealand's islands were designated as nature sanctuaries. One hundred years later, numerous ecological restoration programs were actively underway on near-shore islands in New Zealand. According to Bade, it is clear that islands have played a central role in natural heritage conservation in New Zealand – their separate, bounded, and isolated characteristics have made them appealing as places where native flora and fauna can be protected from human activity on the mainland – but Bade asks what have such features meant for the cultural heritage on the islands? To consider this question, Bade explores natural heritage conservation and the conservation of cultural heritage using two island case studies in Auckland's Hauraki Gulf. Both Rangitoto and Motutapu islands have a history of Māori occupation and settlement, as well as of early European industry and farming, and military use during World War II. Rangitoto was designated as a reserve in 1890, and is largely considered a 'place of nature' due to its volcanic appearance and extensive forest, while Motutapu is largely farmland and undergoing ecological restoration to become 'natural'. By means of such focus, this chapter draws on ideas of 'nature' and 'culture' as key geographical concepts, and explores the influence of islandness on heritage management.

Further comparative work on another set of challenges for island peoples is undertaken in Chapter 7 by Russell Fielding. This chapter first examines the waste-to-energy facility currently in operation on the Caribbean island of St Barthélemy, which takes in municipal solid waste as its fuel source, and then provides thermal energy for the desalination of seawater for municipal and industrial use. Special attention is paid to the history, process, public reception, and challenges to the facility's success, and then to the prospects of the waste-to-energy process on St Barth. Comparison is then made to the stalled proposal to install a similar facility on the island of St Croix in the United States Virgin Islands. By means of comparative analysis, the chapter illustrates how human geography works at the interface of environmental management to deepen insights about island places and island societies, economies, and polities. Adeptly, Fielding establishes that the different outcomes in waste-to-energy applications on the two islands are primarily due not to technical impediments but rather to variations in culture, economics, and political histories.

The Caribbean – and Barbados in particular – is the focus of Jon Pugh's work in Chapter 8, wherein he, like Karides, is concerned to excavate a range of subaltern voices as part of a larger project to address the profound problems in postcolonial thought that are associated with feeling what he calls the history, pressure, and weight of one's own words. Pugh's focus is upon certain tropes – mimicry and mockery, inventiveness, and suspension and impasse; these concern him as he seeks to understand the struggle for postcolonial island cultures to articulate life 'on their terms'. The chapter first considers two essays by St Lucian poet and writer Derek Walcott and then draws on empirical research on the Barbados' Landships that, over decades, have provided welfare and social and organisational structure for the poor, and which represent a network of grassroots and collectivist organisations enabling political organisation and development. In drawing upon Walcott and the Landship, Pugh is able to examine the limits of articulated language and the power of different kinds of silence and of moments of artificiality that are elemental in the postcolonial condition, which reveals how, as Pugh notes, 'the naming of practices, things, rocks, rivers, forests, and other objects feels borrowed and not owned'.

I have also wanted to produce a collection that is more than the parts, and therefore invited my co-authors to read each other's penultimate drafts, generate questions from their engagements with the text, and then participate in a group conversation to explore those questions. That conversation is recorded in Chapter 9, and airs a number of issues with which island studies scholars continue to wrangle, not least in relation to geographical knowledge-production and the practice of geography, for example in the field. Attention was paid to their motivations and values; their epistemic frames; their understandings of the relationships between human geography and island studies;

and the ways in which all of these affect their practices in the field and other settings. The chapter serves a critical function often missing from edited collections: namely to synthesise across the disparate works and find points of commonality that have wider reach and greater salience to researchers, teachers, and students. The aim of such collaboration and ensuing conversational labours was to invite contributors and readers to peer into apertures rather than read this project as an opportunistic collection of papers bookended by an introduction and conclusion (see also Stratford, 1999, which features short interviews between editor and author before each chapter).

Finally, Chapter 10 is penned by Stephen Royle, and was conceived on the island of Ireland in the final months of his role as Professor of Island Geography (Queen's University Belfast) and then completed in the Japanese archipelago in the last weeks of 2015. This final chapter is meant to further extend the integrative characteristics of the collection. It required Royle to engage with Chapters 2 to 9, and use them to reflect on where island geography has come from, and where it might lead. In timely fashion, Royle underscores the need for circumspection – the future is hard to predict; and he emphasises the imperative that whatever we do, we do so with a critical eye, ever alert to discourses and practices that continue to marginalise other-than-continental and metropolitan modes of understanding. And he reminds us of the need to consider, always, the dynamic and productive interplay among sea, island, territory, place, space, and connection. I hope that this book reflects a careful observance of such advice. I might be pushing the metaphor here, but let this work be seen as an archipelago of geographical reflections on islands.

2 The deep Pacific

Island governance and seabed mineral development

Katherine Genevieve Sammler

Introduction

The ocean has long been imagined by those in many cultures as a wild frontier, an alien place. Even more so, the deep sea has been understood as a place of profound mysteries, obscured from extensive scientific inquiry by a few thousand metres of dark, cold, salty seawater. As the place where life originated and evolved on this planet, the deep sea holds clues to our evolutionary past in its myriad diverse, extra-terrestrial, aquatic life forms. Very little of the ocean's depths has been explored in detail because of its immensity and the expense of operating in its extreme conditions. Yet, in the last few years, global metal markets and technological advancements have made deep seabed resource extraction increasingly feasible. Seabed mining remains an experimental technology intended to dredge up mineral sands, nodules, cones, and crusts that contain ores accumulated on and within the seafloor. In the struggle to meet resource production demands, some have turned offshore to satisfy growing food, fuel, and mineral appetites, implicating the seabed as the latest location for large-scale resource extraction. However, arguments put forth regarding seabed mining as inevitable, as essential progress towards harnessing this 'frontier' space, disregard previous attempts and failures of such endeavours.

While relatively little is known about the deep sea compared to land, data are being collected that fundamentally change how ocean spaces are thought about and engaged with. Beneath the surface of our planet's expansive oceans lies the prospect of a new gold rush (after Shukman, 2013), one in which, in pursuit of sunken geologic treasures, companies scour the seabed for minerals and rare metals. The desire to locate such deposits is spurring a surge in attempts to map sections of seabed in high resolution, and to probe the ocean floor to determine the extent and composition of its raw materials. These technical assessments are steps in a process to assert national marine territorial boundaries, key to transforming the seabed into a space for industrial-scale resource extraction.

Before mining has even begun, one effect of proposed projects to mine the seabed is the constitution of this vast and historically invisible resource space as a location of desire, hope, friction, and anxiety – with important implications for evolving state practices of resource governance. A sense of urgency then characterises attempts to exploit this seafloor wealth, and it is evident in discourses of resource scarcity and narratives about the inevitability of development in the deep. In response, Pacific peoples, for example, are grappling with how to determine new legislation, environmental management regimes, and economic benefits associated with such a space.

This chapter aligns itself with a 'focus on Oceanic Studies to reorient our critical perspective from the vantage point of the sea' (Blum, 2013, p. 155), and with the work of various island studies scholars. The aims are to present an historical analysis of the shifting geopolitical and economic conditions that have produced the ocean as a space of industrial development and, in doing so, to focus on ocean materiality and notions of territory. Drawing on insights from the scholarship of political geography and islands studies, my intention is to investigate why and how island locations are central to the development of proposed seabed mining activities, and to ask what makes this moment unique in the record of these activities and allied industries. These labours are deepened and grounded by reference to an empirical case study of New Zealand, where the national government has made recent forays into promoting and regulating seabed mining.

Such work is significant because it is the first investigation into the emerging practices of national territory and sovereignty on the seafloor, contributing to understandings of the continued political expansion of islands beyond the land/sea binary. It is motivated by emerging spatial theorisations of the vertical dimension that are asking us to move away from thinking of territory as two-dimensional planar areas to the exclusion of activities based off-land. The chapter presents an investigation of the processes that construct ocean/island/seafloor imaginaries and of their co-constituted material assemblages – this in order to think beyond islands and oceans to analyse the seabed by considering its legal construction as the prolongation of the nationally bounded continents and simultaneous manifestation as a contingent and shifting submerged space. Thus, recognising the powerful material realities of the ocean, this research will contribute to human geography and island studies by investigating the spatial ontologies of ocean spaces.

To such ends, what follows are sections on the creation of the legal oceanscapes, the relationship of islands to this changing ocean space, and a focus on New Zealand's approach to legislating and regulating seabed mining within its marine territory. The historical context offered in the first section offers insight into the Law of the Sea as it relates to the development of the seabed mining industry and more broadly the capture of ocean space within the nation-state system, utilising the juridical invention designated as the Exclusive Economic Zone (EEZ). In the second section, analysis focuses on how jurisdictions are inscribed onto island marine spaces – places where the sea is integral to identities and livelihoods, where land and sea distinctions dissolve, or never existed. As one of the first governments to pass legislation regulating seabed mining, New Zealand is a critical exemplar to draw on to analyse the applied rendering and regulation of the seabed, on which many island governments, industry investors, and environmental activists keep watch. Two proposed New Zealand mining projects will offer insight into the process of regulating industry seeking to operate on the seafloor.

A brief history of offshore mineral desires

> The desire for seabed transformation from *aqua incognita* to productive metallurgist, situated on the perimeter of the social – a frontier – required a decoding and recoding of flows and permutations, of definitions and practices, of land/sea territorial boundaries.
>
> (Deleuze and Guattari, 1983, pp. 175–6)

Contemporary relationships between national land and sea territories have evolved over many decades. It is useful to understand the shifting historical ideas and conditions leading to the current situation in order to contextualise the social construction of ocean space in relation to the legal and technological 'taming' of wild seas (Steinberg, 2001). Desire both to develop and protect ocean space manifested in the 1982 signing of the third United Nations Conference on the Law of the Sea (hereafter, the Law of the Sea or UNCLOS). The treaty divided up the world's largest resource space – some 361 million square kilometres (139 million square miles) – three times the area of all the continents combined. Among the several motivations informing UNCLOS was the desire to create a stable legal framework to extract minerals from the seafloor.

Fast-space and loose-space

Prior to the United Nations' formalising of the Law of the Sea, ocean space had been subsumed under various arrangements of exclusive and inclusive uses. For example, in an attempt to monopolise trade routes and exploratory voyages, in 1494 those who drafted the Treaty of Tordesillas partitioned the known maritime world between Spain and Portugal. Yet just over a hundred years later, in 1608, such exclusionary practices gave way to *Mare Liberum*, a legal precept invoking the freedom of the seas and sanctioning all nations' rights to navigation. In the early eighteenth century those seas, common to all, again were claimed as territorial waters. Many coastal nations extended their sovereignty offshore to three nautical miles (5.6 kilometres), a distance supposedly based on the range of a shore-based cannon. After a 1945 presidential proclamation expanded the United States' sovereignty past territorial waters to an unspecified distance of the continental shelf (Truman Proclamation, 1945), many national governments began staking claims to marine territories of varied distances offshore. Each territorialisation represented an attempt to reimagine ocean spaces by superimposing the fixed grid of terrestrial boundaries over the surface of an unruly sea (Deleuze and Guattari, 1987). Steinberg (2013) notes that producing oceanic space in such a manner dismisses the challenging dynamism and turbulent and itinerant fields of ocean materiality. Yet, this imposed static spatiality normalises ocean space in ways redolent of developments in other intractable places – for example, the Arctic. The 'aim of this process is, in the end, to increasingly normalise the Arctic and make it like any other place within the dominant Westphalian system: A region that is governed by states, in which law and order is maintained so as to facilitate investment and commerce without major conflict' (Steinberg *et al.*, 2015, p. 164).

Offshore mineral resources were first uncovered when Her Majesty's Ship *The Challenger* fished manganese nodules from the seabed in the 1870s. The crew's exploration of the deep sea was the first of several expeditions lasting into the 1950s, yet there was little understanding of the ecosystems below. Indeed 'our understanding of the deep ocean was one of low biodiversity, no primary production, no seasonality and a uniformly cold, food-poor, dark, tranquil and invariant environment. It was with this scientific framework that the United Nations Convention on the Law of the Sea (UNCLOS) was written' (Ramirez-Llodra *et al.*, 2011, p. 1). Over time, prevailing understandings about the ocean underwent immense changes as material experiences were transformed by technology. With the decline of sailing and whaling, and the advent of steam ships and improved navigational technology, metaphors, imaginaries, and practices of embattlement and antagonism over fearsome and infinite seas gave way to others invoking paternalistic conservation and calculative management of a bounded and beleaguered global ocean. Current social constructions of oceans maintain a state of flux (Steinberg, 2001). While often represented as a wild place or a last frontier (Glassner, 1991; Ramirez-Llodra *et al.*, 2011), the sea is also known to be greatly affected by activities entangled with modern technologies, such as overfishing, plastic pollution, and acidification (Gregory, 2009; Murray, 2009; Srinivasan, 2012). Even as such activities and impacts move farther offshore and deeper beneath the surface, because the ocean is situated on the perimeter of political and social spheres it remains imagined chiefly as a space outside the human realm (Steinberg, 2001).

While the seabed was once considered a space devoid of life, technological advancements are revealing it to be rich in magnificent amounts of biomass and biodiversity (De Leo *et al.*, 2010). Improved technologies reveal pockets of manganese, cobalt, phosphate, iron, and many other precious ores (Glasby, 2000). In the 1970s, a hundred years after *The Challenger*'s discovery, there was a rush to develop these seabed minerals, but a real economic mandate was never reached and attempts were largely abandoned as metals markets declined. In the three decades that followed, improved metal prices, cheaper underwater technologies, and increased scientific understanding and mapping of the deep sea continued to pique interests in the seafloor as a potential site for mining. However, transforming the seabed from *aqua incognita* to productive metallurgist necessitated new understandings of sovereignty over the seas and re(in)scriptions of geopolitical land/sea territorial boundaries.

Completing this project to normalise the sea necessitated consistent, codified, and internationally recognised jurisdictions. From 1958 to 1982, at the same time as many national governments were engaged in arbitrarily expanding their oceanic claims, delegates at the United Nations' first, second, and third Conferences on the Law of the Sea strived to achieve a consensual standardisation of offshore territorial boundaries that both satisfied desires to secure a framework for resource ownership *and* assured unimpeded navigation. Defining a new jurisdictional category to reconcile these contrasting spatial logics was not straightforward. Instead, it required the ocean to be written into a new paradigm, its flows and permutations to be reimagined – from an untamed frontier to 'an abstract point on a grid, to be developed' (Steinberg, 2001, p. 207). Finally emerging after 35 years of negotiations, the 1982 Law of the Sea captured the oceans in a vision of freely flowing commodities and properly fixed resources.

Exclusive Economic Zones: jurisdictions of convenience

With the writing of new borders across the deep seas, the Law of the Sea apportioned marine spaces by creating new national jurisdictions called Exclusive Economic Zones (EEZs). Only in 1994, after the sixtieth signature was secured, did the Law of the Sea come into force, granting nations control over new and vast marine spaces. As with prior attempts to organise and control these spaces, the reterritorialisation made possible by the Law of the Sea was based on land metrics. Embracing the established and fixed grid coordinate system of terrestrial boundaries, land metrics have no capacity to accommodate the unique aquatic materiality of ocean space. As Hubbard (2013, p. 95) explains in her environmental history of marine space, via technology 'ocean spaces [have] become extensions of the terrestrial spaces dominated by industrialised nation-states'. These territorialisations function through explicit assertions about exclusive access to resources on the seabed and within the water column, while preserving mobilities and flows of commerce. Thus, the EEZ was developed with this particular logic: a strategic jurisdictional ambiguity serving to normalise and make legible the sea (Baldacchino, 2010; Scott, 1999).

The EEZ is the largest national spatio-juridical designation in the Law of the Sea Treaty. Extending 200 nautical miles (370 kilometres) from coastlines, globally EEZs comprise approximately a third of the world's oceans. Coastal states may apply to extend their marine territories even further offshore by submitting for consideration by the United Nations new Extended Continental Shelf (ECS) limits (United Nations, 1982, Article 76). EEZ jurisdictions form a crucial part of the Law of the Sea framework, creating legal precedent for the assertion of resource-based sovereignty – granting sovereign rights over resources, but not conceding sovereignty over the space itself.

> In the [EEZ], the coastal State has: sovereign rights for the purpose of exploring and exploiting, conserving and managing the natural resources, whether living or non-living, of the waters superjacent to the seabed and of the seabed and its subsoil, and with regard to other activities for the economic exploitation and exploration of the zone.
>
> (From Article 56(1a), United Nations Convention on the Law of the Sea © 1982 United Nations. Reprinted with the permission of the United Nations)

Sovereignty is exerted over a resource, not the space containing it, and this distinction enables the legal extraction of minerals and metals from the seafloor without implying the kind of rights or responsibilities indicated by full territorial sovereignty on land, nor offering pragmatic management parameters (Johnston and Saunders, 1987). Framing of the ocean within these jurisdictions creates a space in which nation-states can expand a set of rights framed around specific functional uses. Understanding this mode of organising sovereignty and space offers insights into the political geography of non-terrestrial spaces and territories that are increasingly important as resource prospecting moves further offshore, poleward, and even off-planet (Dickens, 2009; Williams, 2010).

Dominant discourses that narrowly interpret territories as the bounded sovereign spaces of nation-states have been called into question (see, for example, Painter, 2010). In practice, territory manifests in diverse arrangements. Elden (2010, p. 801) argues that territory 'needs to be interrogated in relation to state and space, and that its political aspects need to be understood in an expanded sense of political-legal and political-technical issues'. He suggests that one way to approach territory is to view it as a political technology using techniques to calculate and evaluate land and control terrain. Over time, desired objects of calculation and control have changed and new political technologies have emerged. For example, to command submerged terrain, the EEZ was created as a strategic political technology, enrolled to normalise ocean space and secure it for national endeavours; it is the materialisation of a specific territorial imaginary still in the process of being produced technically, scientifically, and practically.

Arguably, the inchoate status of the EEZ requires further theoretical and empirical investigation. Drawing on intellectual contributions from political geography and island studies, I argue that the EEZ should be investigated as a novel geographical strategy optimally understood by reference to its historical context and the material conditions under which it was and is being produced (Baldacchino, 2010; Elden, 2007, 2010; Steinberg, 2005, 2013).

Island/ocean geographies

> ... the idea of smallness is relative; it depends on what is included and excluded in any calculation of size. [Among those who hail from continents] their calculation is based entirely on the extent of the land surfaces they see.
>
> (Hau'ofa, 1994, p. 152)

As an experimental political technology, the EEZ is being implemented within an existing assemblage of cross-scalar relationships that draw into a complex web island nations, subnational island jurisdictions, metropolitan powers, intergovernmental organisations, and global capital. Economic drivers coalesce with political will and technological capability to enable certain actors to press the deep sea into industrial service, and this context suggests the particularity of the events now informing and constituting seabed mining.

Seabed mining is arguably a distinct *moment* in the accretion of conditions and constraints that characterise both ocean imaginaries and materialities. In such light, understanding territory as a produced phenomenon demands an investigation into the historic, economic, legal, and technical facets involved in the political reorganisation of ocean space (Elden, 2010). The land/sea binary dissolves into a constellation of island-island, centre-periphery, developed-developing, surface-volume, fixed-flowing relations converging in material ocean spaces (Pugh, 2013). As ocean territories are operationalised in distinct ways across island nations in the Pacific Ocean, for example, the scales of national and international maintain some meaning as categories of analysis, yet the ocean itself exceeds such human spatial and temporal scales. Certainly, colonial and postcolonial relations between islands and continents contribute to this new territorial category being superimposed over existing embedded institutions, infrastructures, and congealed relations of power in the Pacific; so, too, do connections between island nations, islanders, and oceans.

Island constellations / ocean relations

At various times, European and American rhetorics have portrayed the Pacific as an ocean void scattered with empty islands (Matsuda, 2007). Discursively constituted as vacant space, the Pacific became a ready target of exploitative practices of resource extraction. Legacies of these colonial practices still scar island landscapes and oceanscapes. Passed from one colonising power to another, many Pacific island territories have suffered extensive resource extraction – for example, in the form of fishing, agricultural cash crops, timber harvest, guano extraction, and terrestrial mining (Burnett, 2005). Jurisdictional arrangements, originating from imperial interventions in the Pacific, contribute to heterogeneous political, economic, and social topologies. The results are varied compositions of sovereignty, territory, jurisdiction, and enclave across a spectrum of potential political manifestations: 'a constellation of subnational island jurisdictions ... illustrating the flexibility and tenacity of global capital, of federal politics, of smaller autonomous territories, as well as of sovereignty and the geography of power generally' (Baldacchino, 2010, p. xxii).

Many dependencies and associations thus exist within and between islands, or between islands and their former or current continental powers. In the Pacific, examples of these relational archipelagic arrangements include the Cook Islands, Niue, and Tokelau. Passing from British control, these islands are now in 'free association' with New Zealand on matters of mutual interest – allowing their peoples citizenship, budgetary assistance and diplomatic representation, and varying degree of independence. These island groups, together with New Zealand's unrecognised Antarctic claim, comprise The Realm of New Zealand constituted under the nominal head of state, HM Queen Elizabeth II.

There are, of course, many other relational webs that bind together island governments in archipelagic interconnectivity. Kiribati and Samoa are supplied defence assistance by the governments of Australia and New Zealand, having no military forces of their own. Search and rescue services are provided by New Zealand 'from the Equator to the Antarctic, and from half way to Australia to half way to Chile', one of the largest in the world at 30 million square kilometres (11.58 million square miles) (Maritime New Zealand, n.d.). These examples demonstrate some of the strong political, economic, and strategic connective dynamics in existing Pacific relations.

Beyond these connections at the level of the nation-state, island peoples share associations that transcend official borders and exceed land/sea divisions. Indigenous populations with strong historical claims to the sea have attempted to call attention to the extreme biases of the categorisation and overlay of linear boundaries over a lived and living ocean space (Stratford *et al.*, 2011). Those in subnational and cross-national networks of Pacific island peoples negotiate various island jurisdictions, asserting their rights to fisheries and marine spaces of all types, maintaining independent identities, politics, and sovereignty movements (Hau'ofa, 1994). In New Zealand, several Māori *iwi* (tribes) continue to make claims on the coastal seafloor and seabed with the conviction that their treaty rights apply to 'dry land' and to the marine area as well (Douglas, 2010). Decidedly, these indigenous discourses of land/sea spatiality differ from coloniser practices of political boundaries; island cosmologies residing as much in the sea as on land (Hau'ofa, 1998).

Submerged terrain/submarine materialities

The sorts of tensions that typify debates about seabed rights as they manifest in New Zealand are neither surprising nor singular. Indeed, Pacific nations are presumed to have the most to gain from developing their seabed resources because their EEZs are tens, hundreds, sometimes even thousands of times

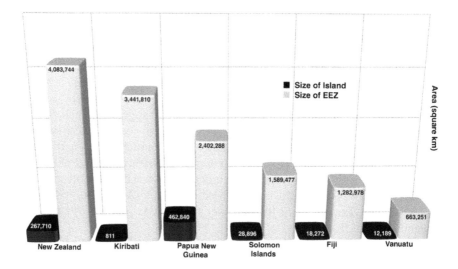

Figure 2.1 Select Pacific Island nations: land and territorial areas. (Extended Continental Shelf areas excluded, as they continue to be assessed by UN.) Image by Katherine Sammler. Data sources: United Nations, 2015.

larger than the island groups' landmasses (Figure 2.1). Even the smallest island nation may be granted 125,700 nautical square miles (431,000 square kilometres) of marine territory – estimated on the area of a circle (πr^2), where r is 200 nautical miles, assuming island circumference of one. Of course, many of these islands exist in close enough proximity to have overlapping claims for EEZs, which must be split among them; requiring precise calculations to delimit shared marine borders.

Beyond considerations of size, the Pacific region is expected to contain an amount of minerals greater than that found in other ocean basins (Boschen *et al.*, 2013). Many of these small nations are situated above geologic conditions ideal for the formation of seabed minerals: converging and diverging tectonic plates with continental and oceanic crusts mashing together or oozing apart (Pandey, 2013). Thus, both islands' geographic position atop *terra mercurial* and the tremendous size of their EEZs offer ideal material conditions for those wishing to access extensive seafloor resources. Mark Brown, finance minister for the Cook Islands, contends that mining in its EEZ could transform the nation into one of the richest in the world in terms of per capita income and 'increase gross domestic product a hundredfold' (in Neate, 2013, n.p.). As a governmental entity, the Cook Islands is currently ranked in the bottom ten nations for national purchasing power parity (PPP), a statistic representative of many Pacific Small Island Developing States (SIDS) (Hegarty and Tryon, 2013). Ostensibly, this position is due to a lack of natural terrestrial resources and isolation from foreign markets, and certainly it is a product and/or legacy of colonial and postcolonial extractive industries, neo-colonial free trade agreements, and debtscapes.

Extensive resources within these submerged territories indeed have the gleam of sunken treasures for developing nations and are economic lifelines. For this reason, even though management strategies are emergent and extraction techniques untested in the deep sea, many SIDS governments and global investors are eager to start mining. In fact, some Pacific island governments are at the forefront of national seabed development. Yet, the EEZ is being implemented differently across the Pacific's national governments, in part as a function of existing and emergent governance structures, and significant political challenges and incalculable environmental risks stand in the way of mining seafloor riches. Island nation governments and enterprises seeking to engage in this industry are still creating legal frameworks and generating governance and other capacities to govern the extraction of non-living resources from their expansive marine territories. Notwithstanding such complications, various seabed mining projects are in the planning stages across the Pacific. For example, national governments in Papua New Guinea and New Zealand have taken steps to implement legislation and engage in economic calculations that would permit their first mining operations. As a result, and noting that the governments of Australia's Northern Territory and of the nation of Vanuatu have issued moratoria on mining in waters off their coasts, some island nations may now be on the brink of significant economic, ecological, and governance transformations.

It is important to consider that EEZ boundaries were created in tension with the context of ocean physical extremes, among them pressure, temperature, and darkness. A corrosive fluid of unbounded flows and mobilities (Steinberg, 2013), the ocean remains an extremely difficult space upon which to impose governance regimes or to enforce regulations. Organising the vast horizontal and vertical spaces of the ocean has proven challenging to institutions historically developed to govern, manage, and develop land (Palan, 2003). The wayward physical properties of the ocean restrict neither ecosystems nor pollutants from spilling over politically and legally constructed boundaries. Laws and policies are not so easily inscribed on such a place (Scott, 2010). Thus, the materiality of the ocean plays an active role in its constant reformulation as it generates waves, exerts buoyancy, absorbs light, transports heat, and dissolves materials which respectively limit infrastructures, afford navigation, conceal objects, propagate energy, and corrode solids (Lehman, 2012).

Within this tempestuous medium, seabed mining is an untested industry necessitating the extrapolation of land-based strategies of management and excavation to locations beneath oceans. Sediment plumes from mining operations and post-extraction mine tailings returned to the seabed are thought to have devastating potential. Suspended sediment may increase toxicity or block out sunlight in the water column, and clog the filter-feeding apparatuses of benthic organisms. Plumes that block sunlight would decimate photosynthetic plankton, the foundation of the aquatic food chain, posing a risk to ocean ecosystems and the communities that depend on them. These risks are particularly dangerous to Pacific nations, due to their exceptional dependence on marine spaces economically and culturally.

Topologies and contiguities

Although seabed mining has the potential to create many political, economic, social, and ecologic changes, most Pacific island governments are still in the early stages of devising legislative and administrative programs that are based on the understanding that the EEZ is a territorial category. Practically, this work involves several difficult tasks – producing codified rules and norms, and converting these into legal and regulatory performances to bring this new national space into being. As this space emerges and becomes available for exploitation, extraction, and accumulation, the boundaries drawn around islands in the Pacific represent new potential island-seabed topologies of power, drawing attention to the already 'complex and overlapping intermediations between islands and between land and sea' (McMahon, 2013, p. 56).

Previously separated by hundreds of kilometres of ocean, island nations now have territorial contiguity (Figure 2.2). By creating such maritime border connectivity between Pacific nations through their adjoining seabeds, additional topologies prompt questions about access, flows, cooperation, and regional governance. Historically, lines drawn on maps of the ocean could be considered lines of division or of connection (Steinberg, 1999), and EEZ boundaries can be represented in both ways. Globally, national governments now have stewardship over almost 40 per cent of the Earth's ocean area (131 million square kilometres/50.5 million square miles, which excludes Extended Continental Shelf claims; see UNEP/GRID-Arendal, 2011). Thus, they are able to protect and/or exploit resources within the water column and on the ocean floor. Yet, with a large portion of the west Pacific now a mixture of EEZ jurisdictions, it is difficult to imagine how watery volumes of flow and movement, interlaced with these leaking borders, will be managed at, or exceeding, the national scale. Governance within the region is not homogeneous but, rather, functions within networks of power relations among those actors and entities implicated in national, regional, and international governance, and in various flows of capital. Indeed, capitalism 'decodes and deterritorializes with all its might' (Deleuze and Guattari, 1983, p. 336) – it penetrates the seas, as the seas transgress the borders, and its protagonists seem to see only the land and resource beneath.

Along with the general difficulty of managing ocean space at all, varying colonial and postcolonial pasts and presents, and the numerous sub- and transnational archipelagic arrangements and relationships point to the EEZ being implemented very differently in practice across the Pacific. Notwithstanding, regional governance associations influence island-to-island knowledge creation and dissemination, and manifest as substantial instruments of power.

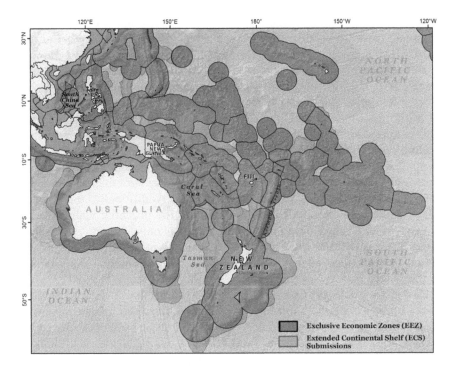

Figure 2.2 Territorial contiguity in the Pacific created by EEZ jurisdictional borders. Image by Katherine Sammler. Data sources: VLIZ, 2014.

Island studies literatures highlight how a territorial border defined by a land/water interface is 'critical feature of islands but by no means is it definitive, for the land and sea boundary is a shifting, fractal and paradoxical one' (Stratford *et al*., 2011, p. 115). As island territories have thrust seaward, past land-sea boundaries and into the deep, the annihilation of such binary distinctions is exposed by the existence of the EEZ.

Furthermore, meditation on archipelagic thinking reorients island inquiry away from a focus on connections to a mainland. Instead, archipelagic thinking considers the strengths of island-island connections, 'relations that may embrace equivalence, mutual relation and difference in signification' (Stratford *et al*., 2011, p. 113). In this conceptualisation, because of island nations' interconnectedness and cooperation they may be better positioned to manage the seaward thrust of their territorial boundaries. Yet, competition over mining contracts could motivate reduced regulation or decreased royalties in a race to the bottom. If mineral prospects are pursued near a shared EEZ border, transboundary issues of mobile pollutants or sprawling ecosystem damage could become a point of conflict.

Despite the potential for conflict, cooperation is not new among Pacific island states. The Secretariat of the Pacific Community (SPC), formally entitled the Pacific Commission, is a remnant of colonial powers in the region. Its membership, when founded in 1947, comprised territorial administrators from Australia, France, New Zealand, the Netherlands, the United Kingdom, and the United States. As island nations gained independence or self-government in the 1960s and 1970s, they became members of the SPC under their own political leadership. In many technical, economic, and political matters, the organisation now represents 22 island nations and their metropolitan counterparts. For example, the SPC is amassing technical information under its Applied Geoscience and Technology Division. With funding from the European Union, the SPC Deep Sea Mining project is establishing recommendations and best practices for formulating legislative frameworks and environmental management in anticipation of seabed development. Additionally, the Pacific Island Forum has released several reports on the potential viability and drawbacks of seabed development for its 16 member states. Such relational webs constitute multi-directional flows of information and power among islands, now connected by the political and physical topography of the seabed.

Despite the existence of both shared regional and global governance frameworks on which Pacific governments may draw, there are many competing ideas regarding deep seabed mining. Those working in development groups express hope that funds flowing from seabed mining activities will provide their nations with much-needed income to improve infrastructure and spark further investment. In contrast, local environmental groups and sceptical citizens envision seabed development as yet another transfer of resources from their islands to a mainland or continental force, leaving behind environmental ruin unlikely to be fully remediated.

The geopolitics of New Zealand's deep sea mining strategy

> May our mountains ever be / Freedom's ramparts on the sea, / Make us faithful unto Thee ...
>
> (*God Defend New Zealand/Aotearoa*,
> New Zealand's national anthem)

New Zealand is a developed economy and archipelagic nation in the Pacific region comprising two large islands and over 700 smaller offshore islands. It is important to deep sea mining politics because of the national government's pioneering efforts in the development of EEZ legislation, policy, and management structures. Focusing here on New Zealand enables consideration of what it means to envisage and implement marine spaces, which are especially difficult to govern – both alluding to and evading categorisation as national sovereign territory. Such consideration, in turn, offers important insights into emerging legal, political, economic, and social constructions of the deep sea.

I undertook three months of fieldwork in New Zealand in 2014, work that resulted in many lengthy interviews with key stakeholders from regulatory offices, non-government organisations, grassroots environmental movements, mining companies, and legal and Pacific scholars. I consulted documents in the New Zealand National Archives in Wellington, via multiple online government sources and, later, did the same in the United Nations Archives in New York City.

Thinking about New Zealand's seabed mining development, in what follows I argue that two characteristics dominate 'the conversation'. First is a persistent tension between rhetorics and practices, imaginaries and materialities. Documents reveal, and stakeholders speak of, the development of the seabed as 'inevitable', even as they acknowledge that economic and ecological uncertainty greatly endanger the feasibility of undersea mining projects. Second, as revealed by documents and stakeholders, there exists a patchwork of legislation in response to the EEZ territory, which they understand as a deeply ambiguous jurisdiction.

Land of the long white cloud, sea of the deep crown minerals

Having among the largest marine territories in the world, New Zealand's more than four million square kilometres (1.54 million square miles) of EEZ holdings are roughly 20 times the size of its landmass (EPA, n.d.; Figure 2.3). New Zealand is an influential member of regional inter-governmental organisations, and maintains various arrangements and associations with multiple islands nations and territories. Its relative dominance in the relational topology of Pacific seabed development makes it an important barometer in the region, as its development and security activities will be significant for those in many Pacific nations.

At time of writing, New Zealand differs from many Pacific governments currently working to introduce legislation on seabed mining. It is a much larger island group with a more diverse economy, more resources overall, and a greater capacity to initiate novel governance frameworks for its maritime territories. New Zealand declared 200 nm of marine territory in 1978, at the same time as many other nations were scrambling to assert strategic claims offshore. Its claims included the Kermadec, Chatham, and several sub-Antarctic islands.

In 1991, the Crown Minerals Act staked New Zealand's claim to the entire continental shelf beyond the EEZ for the mining of minerals, petroleum, and other natural resources, with rights to explore and exploit these resources vested in the Crown. The national government ratified the Law of the Sea treaty in 1996, but only on 28 June 2013 did the Exclusive Economic Zone and Continental Shelf Act 2012 (EEZ Act) come into force. The New Zealand Environmental Protection Authority (EPA) began evaluating its first marine consent application in early November 2013.

Conservation groups, fishers, and other concerned citizens organised a campaign and protests against this instantiation of seabed development

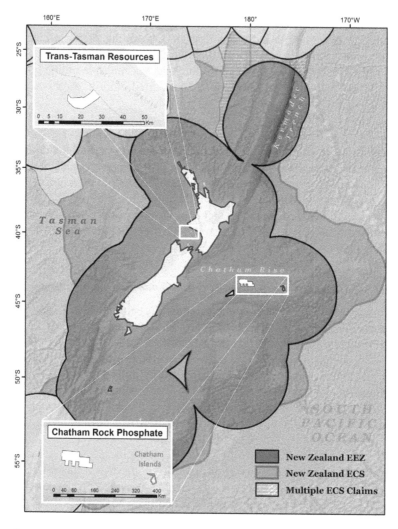

Figure 2.3 Proposed mining sites, New Zealand's EEZ and ECS marine territory. (Sites pertain to the Trans-Tasman and Chatham Rock Phosphate companies.) Image by Katherine Sammler. Data sources: VLIZ, 2014; EPA, 2014; 2015.

because of concerns about the experimental status of mining technologies, and perceived lack of oversight and economic fairness. The national government's response has been to amend the Crown Minerals Act 1991 to create exclusion zones around prospecting, exploration, and extraction vessels and structures. The Act authorises the New Zealand Defence Force to arrest and detain anti-mining protesters, and enables steep fines and potential incarceration (Crown Minerals Amendment Act 2013, 101B-C). The amendment instigated its own protests and a tirade from Moana Mackey of the New Zealand Labour Party:

> [T]he United Nations convention on the continental shelf tells us that we do not have the right to pass legislation on the water column above the continental shelf. We have the right to exploit minerals in the continental shelf, we can drill into it, and we can tunnel into it, but in the water we have no jurisdiction ... How is it, then, that we have the right to regulate shipping activity that is protesting in the water above the continental shelf? We do not. It is humiliating that the Government is sitting here passing law that we have no right to pass. You might as well pass a law on Mars, because you do not have jurisdiction there, either.
>
> (New Zealand Parliament, 2013a)

The debate over seabed mining off the North Island of New Zealand clearly demonstrates the contradictory imaginaries that have produced ocean territories. In tension here are complicated dual modes to secure resource ownership and ensure the freedom of navigation. Restricting offshore activity serves to privatise ocean space to approved commercial entities and activities by removing protesters. Providing the legislative certainty for offshore investors guarantees the private rights of the extractive industry over a public commodity represented by the Crown. This situation highlights the jurisdictional ambiguity and patchwork legislation being passed to fill regulatory gaps over the EEZ. It also reveals both environmental and economic anxieties produced by potential seabed mining projects affecting New Zealand's maritime territories.

El Dorado meets the white whale: inevitability vs uncertainty

It is postulated here that seabed mining ventures are typified by visions of El Dorado: tempting riches with a high barrier to overcome, an inevitable promise with the elusiveness of the white whale. Nevertheless, stakeholders in New Zealand's seabed mining debate appear to accept, whether reluctantly or eagerly, that exploration and extraction are inevitable – next steps in the unrelenting march of progress. Comparing onshore and offshore mining, in September 2014 a geologist shared with me the following:

> Either way, it's gonna happen, so which would you rather? Which has greater effects? Which is more easy to mitigate? People don't think about this stuff. They just say 'not in my backyard' or 'shut down all mining.' What you mean 'shut down all mining'? Life as you know it would cease to exist. Everything that you do – your cell phone, your TV, your car, the plane that got you here – everything requires metals.

In fact, this period is not the first in which seabed mining has been a fervent topic of speculation and debate. Motivations driving the push for many national governments to formalise the Law of the Sea were prompted by predictions of global mineral shortages and bold faith in the potential for riches from the seafloor. Yet, international research and development investments made through the 1970s and 1980s delivered little return (Glasby, 2000). When the Law of the Sea came into force in the 1990s, seabed mining again promised riches that never materialised because of low market prices for metal and large upfront capital costs. Yet, the uncertain economic benefits of seabed mining to New Zealand are only partly articulated in varied campaigns launched in objection to the industry (for example, consider Deep Sea Mining Campaign, Greenpeace International, Kiwis Against Seabed Mining). Most public discussions about uncertainty focus not on economic matters but on a lack of scientific and environmental data, a prominent and valid concern made stronger by a range of indeterminate economic factors, as discussed below.

Both in area and depth, the scale of EEZs is known to pose immense difficulties to data collection and modelling. The seafloor challenges technical abilities because it is an environment of extremes: freezing temperatures, zero light, corrosive salinity, and crushing pressures. One of the barriers facing the seabed mining industry in New Zealand is the investment needed for data collection at a potential mine site. The viability of the site as well as the permit application necessitates prospecting and exploration, processes that include gathering precise technical information about oceanographic conditions and benthic habitats, taking core samples, and doing economic and environment modelling, the costs of which are borne by the applicant. These data are proprietary and pertain only to the area of interest. In June 2014, an environmental economist explained to me that one problem with this situation is that we 'need to know about places that don't have minerals, not just the places that do, to determine the relative ecological importance'. Mounting arguments against mining emphasise the point that not enough is known about the ecology of prospective mining areas to create a baseline from which to measure mining impacts and ecosystem recovery. Beyond the damage done by physically removing the rock, which is known to completely change the immediate vicinity, there is much uncertainty and concern about the larger impacts of sediment plumes or food chain reactions. Thus, in August 2014 one grassroots organiser shared with me his concern about how 'the uncertainty of the uncertainty is uncertain'. He worried about how little is understood about the interrelationships, interconnections, and interdependencies of ecosystems and physical systems off the coast of his home in New Zealand, especially since there is no period of baseline data with which to compare the impacts of proposed activities.

Illustrative of these tensions is the Trans-Tasman Resources Limited (TTR) mining proposal. Submitted in October 2013, the permit sought to mine iron sand off the west coast of New Zealand's North Island (broadly 39.6° S, 174.1° E). It was the first to go through the marine consent process established by the New Zealand government and administered by its EPA. This operation was to take place 12–19 nm (22–36 kilometres) offshore, in water 19–42 metres deep, which is relatively shallow when compared to most sites of offshore mineral resource evaluations (TTR, 2013). The project proposed the excavation of up to 27 million cubic metres of seabed sand per year in order to extract its iron content. The processing of the sand would have taken place on-board ship with over 90 per cent of the sand then being returned to the seabed. The iron ore concentrate would be available for direct export. Despite the tax and royalty payments to the New Zealand Government, predicted to be in the range of NZ$50 million per year (US$38 million), such estimates were criticised for disregarding the negative economic impacts on tourism and fisheries' interests. For example, it was noted that 'to assess the true economic impact of the project, potential negative effects as well as input assumptions such as iron ore prices need to be included into the specifications of the model' (EPA, 2014, pp. 155–6). A New Zealand regulatory agency staff member told me in July 2014 that:

> the economic arguments are very thin. And they're thin because people don't understand what the inputs are. And if they begin to understand the inputs, it's being played out on an environmental basis very poorly understood … we don't understand where the concentrations of fish are, we don't understand migratory patterns, we don't understand nurseries, we don't understand the food chain … You interrupt that chain at any point … it will have an economic consequence.

In fact, in June 2014 TTR's marine permit was declined. A decision-making committee appointed by the EPA reported its main concerns – including environmental impact uncertainty and lack of clarity about the extent of economic benefit to New Zealand (EPA, 2014). Indeed, the EEZ Act states that 'If, in relation to making a decision under this Act, the information available is uncertain or inadequate, the EPA must favour caution and environmental protection' (s.61(2)). Thus, and given fluctuating commodity markets, it has yet to be proven that industrial-scale seabed mining can be done safely in terms of the principle of the precautionary principle enshrined in the Act, or done profitably (Stegen, 2015). In addition, improved offshore mining technology will be in competition with improvements in onshore mining technology and efficiency gains in the latter are likely to negatively affect the former. This precautionary principle places a burden on seabed activities more onerous than that exacted upon land-based pursuits, and is based on an understanding that there are insufficient data available in relation to mining in ocean environments.

Building the plane while flying it: the production of patchwork legislation

In anticipation of seabed mining pursuits, to allocate uses and minimise conflicts between ocean users in June 2008 the New Zealand Government hastened to draft legislation and implement management regimes within the EEZ jurisdiction (Agardy, 2010). A patchwork of legislation and management regimes has emerged in stages as opposed to appearing as one comprehensive ocean management framework and allied governance agendas. Many of those with whom I spoke referred to the need to 'fill the gaps' in both governance for, and data collection in, the EEZ. In July 2014, one government policy analyst said that his office was always thinking about the newness of the whole endeavour; a new activity in a new environment by new regulators using new legislation to give permits to new companies. Regulators and industry personnel alike recognise great need for political, economic, environmental, and social data to be collected and analysed in order for this vast region to become a coherently governed national space with the responsibilities that come with bordering other nations' EEZs and the global commons of international waters. The EEZ Act itself is stop-gap legislation, as the Crown Minerals Act 1991 has no environmental impacts regulation. For some, filling these gaps is important for conservation and to develop social licence to operate in ocean spaces; for others, it is important to secure stability for capital investment. Known gaps in legal, environmental, economic, and social knowledge and understanding are being filled in an *ad hoc* manner, and stakeholders have reported a lack of unified vision for ocean management. In parliamentary debates, Moana Mackey (New Zealand Parliament, 2013b, n.p.) more than once has called into question the proliferation of seabed mining legislation, observing that 'we seem to be passing an awful lot of legislation for one industry in this country, to the detriment of the environment and the democratic rights of New Zealanders'.

One of the reasons for this so-called patchwork approach arises from the fact that the Fisheries Act 1996, the Resources Management Act 1991, and other pieces of legislation are now being overwritten by the needs of those seeking to engage in seabed mining. The outcome of this situation is that there are different regulations and regulators for near-shore and farther offshore areas, for living and non-living resources, and for different activities within the same spaces, despite the fact that all of the activities in the ocean will have impacts on each other. In July 2014, this regulatory approach was described to me by a government employee as akin to 'building the plane while flying it'. Within the EEZ there are multiple potential uses, and multiple laws and regulators seem stacked over the top of each other, an outcome partly attributable to the temporal mismatch between planning and development within the EEZ, a lack of vision for what rights and responsibilities are appropriate in these marine spaces, and the vertical potential for numerous overlapping uses. This governance patchwork can only exacerbate the challenges of administering marine spaces, and operating within ocean space.

One example of a remarkable overlap is demonstrated by the second permit application submitted to New Zealand's marine consent processes. In

May 2014, Chatham Rock Phosphate Ltd (CRP) submitted materials to col-lect phosphorite nodules from the seafloor at depths of 250–400 metres in an area called the Chatham Rise, 400 kilometres (216 miles) offshore and east of the South Island (43.7° S, 179.7° E). The proposed project sought to remove roughly 1.5 million tonnes of phosphorite nodules from 30 square kilometres (12 square miles) annually for 35 years, totalling an excavation area of 1,050 square kilometres (405 square miles) (EPA, 2015). All parties acknowledge that breaking apart and suctioning up seafloor materials will destroy all ben-thic organisms in any given mining area since it involves removing the seafloor on or in which benthic flora and fauna live (CRP, 2014). Paradoxically, the proposed mining area overlaps the Chatham Rise Benthic Protection Area (BPA) established under Fisheries (BPAs) Regulations created by the Fisheries Act 1996, and in which trawling has been banned since 2007. In July 2014, one regulator shared with me the following: 'Chatham Rock [mining proposal] is really interesting in a benthic protection area. The benthic protection area was put in place to stop bottom trawling for fishing. So we take that bottom trawl-ing out and we permit people to destroy the seabed by, they call it vacuuming, but I think that's a euphemism. Vacuuming up to a half metre deep.' Indeed, the mechanism planned by CRP for nodule removal is a trailing suction hop-per dredger and drag-head, which can resemble a large, very powerful vacuum.

The BPA regulation specifically states as one of its purposes 'to prohibit the use of a dredge within the benthic protection areas' (Fisheries Regulations, 2007, s.3(c)). This prohibition applies specifically to fishers dragging the bot-tom for shellfish, however, and does not rule out mining. George Clement, the Chair of Seafood New Zealand and chief executive of the Deepwater Group representing fisheries interests in New Zealand, has argued against mining within the BPA, reasoning that the Fisheries Act 1996 was used to create the protection areas because fishing was the only human activity taking place within the EEZ at the time. He commented on this quandary in the follow-ing terms: 'We have an unacceptable situation where the seabed in this BPA can be mined quite legally, even though it is illegal to have a fishing net touch or even go within 50 metres of it ... It's a case of different rules for different users ... the seabed of the BPAs should neither be trawled nor mined' (in Deepwater Group Ltd, 2014, n.p.).

After several delays, in February 2015 the EPA-appointed decision-making committee declined consent. The rejection of the CRP project and of TTR's permit will have significant impact on how people imagine the feasibility of advancing this industry. As such, those in national govern-ments, regional organisations, and corporations are watching develop-ments in New Zealand, as well as across the archipelago of Pacific Island nations, to gauge if there is progress towards politically, economically, and environmentally viable seabed mining. In the meantime, New Zealanders who believe that the legislation failed in its purpose to grant permits and regulate the seabed mining industry are calling for amendments to be made to the EEZ Act.

Conclusion

Over time, the extension of national space into the global ocean has proven challenging to conventional institutions of governance, prompting concerns over economic fairness and environmental impacts. At the same time that the proposed benefits of seabed mining are being secured or delayed by governments, varied conservation groups, fishers, and other concerned citizens are binding together to protest seabed development. Experimental mining technologies, lack of oversight, and economic plunder of this common property are of dire concern to many Pacific peoples, who have long experienced the effects of extractive practices of ocean and island natural resources. Profitability and economic fairness will depend on levels of environmental and economic regulation implemented – both within given nations and across entire regions.

In this light, it is significant that Pacific island nations are heterogeneously interconnected via culture, economics, and governance arrangements, as well as by their recently shared territorial contiguities as formed by abutting EEZs. This archipelagic Pacific must contend with economic markets and drivers, as well as with ecosystems and pollutants that never hesitate to spill over politically and legally constructed boundaries in the ocean. Traits of the ocean's materiality, combined with experimental mining technology, make for a very precarious ecological and political situation. Pacific peoples, while presumed to have the most to gain from developing their seabed resources, bear witness to great uncertainties on the horizon regarding seabed mining practices. Both ocean conservationists and seabed developers interviewed by me expressed concern and frustration over the patchwork legislation and lack of unified vision for ocean policy. The fact that legislation and regulation is still in the process of being written, and that the EEZs of the Pacific are still emerging, make this an important moment for intervention and vision.

While the seabed mining industry has been in a fledgling state for decades, its continued resurgence illustrates the power of the discourses of resource scarcity and inevitability of progress. Assuredly, other Pacific nations will be affected by New Zealand's development aspirations given the ecological, economic, political, and social ties that bind the region together. Possibly they will be hard-pressed to compete for mining contracts. Those on smaller islands face challenges additional to any confronting those who would push for seabed development in New Zealand's EEZ: for example, there are fewer regulators to implement suitable policy and practice, and fewer scientific resources to address uncertainty surrounding marine ecosystem impacts. Moreover, urgent questions remain about who will benefit from seabed mining projects should they proceed.

In this chapter I have advanced the interrogation of territorial orderings of the world via ocean spaces. Recognising that spatial organisations are historically and regionally contingent, EEZs represent one such and particular configuration of space that is both a political technology and a geographical strategy enrolled to secure locations – one that exceeds categories such as national/international or land/sea. Considering the landscapes and seascapes of the Pacific as more than spatial containers where seabed development is emerging, the social, cultural political, and economic aspects of seabed mining reveal the enormous agency of islands and oceans, archipelagos and basins, seafloors and water columns, salt water and sediment. Demonstrating that the historical contingencies of constructing territory are powerfully combined with the constant challenges of operating within the ocean's material conditions, this chapter thus serves to destabilise discourses of the inevitability of seabed development.

3 Islands and lighthouses

A phenomenological geography of Cape Bruny, Tasmania

Thérèse Murray

Introduction

Theories of place are a core concern in human geography. To speak of human beings and their experiences, activities, cultures, and socio-political structures in geographic terms entwines people, locations, time, and environments – as well as the complex interconnections, flows, and interdependencies among them. Place is not an abstraction; places are fundamental to the ways in which we affectively experience those interactions and the world we live in (Tuan, 1979). As Edward Casey (1996) asserts, human beings are 'placelings' in that everything we do, think, say, and experience occurs somewhere and somewhen, in and among places. When we reflect on our own experiences in the world this assertion may seem self-evident. Place, as Tim Cresswell (2004) observes, is a ubiquitous term used in everyday speech. We visit places. We stay and live in places. We talk about places in terms of emotion, memory, desire, identity, power, and change. It is only when we try to articulate just what place means that it becomes clear that this apparently simple and well-understood term is not so transparent after all, and that place can be understood in multiple ways. Understandings of place in human geography are necessarily multiple, varied, and contested, depending on the ontological and theoretical perspective, focus, and purpose of the writer.

Cresswell claims that the dominant approach to theorising place is one that sees place as socially constructed. Such an approach is exemplified by other human geographers such as Doreen Massey (1991) and David Harvey (1993, 1996) who, in very different ways, argue that ideas about how places are constituted, connected, and understood are integral to understanding human interactions, power structures, and inequities. For example, Harvey (1996, p. 261) describes place as a 'carving out [of] permanences' in a world of ever increasing rate of change and loss of clear boundaries. Massey (1991, p. 188) argues for an understanding of place as unbounded and dynamic weavings together of social relations, networks, and flows of power resulting in places that are 'articulated moments in networks of social relations'. Social constructions of place are rich and effective means to understand the dynamics of place. However, there are other ways of knowing. Place can also be understood as having an ontological existential precedence to space, time, social construction, meaning-making, and power relations (Casey, 1996). In this view, human beings are participants in the manifestation of place but not its sole sources, authors, or producers (Ingold, 2000). Human beings and their constructions, both material and social, are intrinsic parts – but not the whole – of the phenomena of place.

This understanding of place is indebted to Martin Heidegger's (1951) work, *Building Dwelling Thinking*. In that account, people are not subjects in a space containing objects. Rather, there is a gathering into a unified whole of finite, time-bound human beings, a bounteous Earth, the changes of the sky and weather, and an ineffable sacredness. Conceived in such fashion, place is dynamic (Hay, 2002). Places 'enfold and are enfolded by other places' (Malpas, 2013, p. 11): they are bounded but not static; nor are they closed off from connections and flows. Such place boundaries should not be understood purely as physical or environmental; they include less tangible horizons created by the geographical imagination, social constructions, representations of the world, memories and emotions, and indeed our mortality and place in time. This material and social boundedness of place provides for the world to appear and be experienced in particular ways. The word 'appear' is used advisedly; to appear means to 'come forth, to become visible' (Harper, n.d. a, n.p.), which implies actuality and potentiality of continuous becoming and fluidity in the manifestation of place.

Heidegger (1951) claimed that the things we construct have a particular role to play in this coming forth of the world into our experiences insofar as they gather the world together according to their essential nature, providing for the appearance of a particular and multifaceted place. Heidegger's example is a description of a bridge that *gathers* the banks of a river together, *permits* both the flow of the river and its crossing at just that place, *stands* in and *bears* the marks of the changing weather, *creates* a cool, dark place of difference beneath and – as is the nature of bridges – *allows* crossings, flows, connections, relations, and exchange among people across different scales, environments, and times. This gathering is more than the sum of its parts. The folding of location, time, and people according to the nature of the bridge forms a whole – a place in a continuous process of becoming and of reconfiguring connections with other places. Heidegger's description of this place is more than a description of what it is and how it comes together; it is evocative, poetic, and conjures a sense of the affective experience of place that, so often, is difficult to articulate.

It was my own emergent and strongly felt experience of a particular place that was the genesis for the enquiry that follows in relation to Cape Bruny and its lighthouse, located at the southern end of Bruny Island, Tasmania (43.3° S, 147.2° E) (Figure 3.1).

I am not an island person, nor do I have a special interest in lighthouses, yet my first visit to Cape Bruny evoked emotional and physical responses, a strong sense of attachment, and a fascination with the lighthouse that has drawn me

Figure 3.1 Cape Bruny Lighthouse. Photograph by Thérèse Murray.

to return again and again. It was useful to think about Cape Bruny and its lighthouse in terms of social constructions – island and lighthouse representations, history, and political and mercantile roles. However, those labours did not fully explain my initial and ongoing response to this place, a reaction that preceded an in-depth knowledge of the site's stories and functions. This disjuncture between the felt and thought prompted a series of questions about what knowledge might be accessible through such experience. In that inquiry, Heidegger's bridge came to suggest the possibility that the lighthouse may play a central role in the investigation that has ensued. This chapter was motivated by such musings and provides a first-person, inductive, phenomenological, and haptic consideration of the place at Cape Bruny and role of the lighthouse in gathering the sea, (is)land, sky, and people in the phenomena of a particularly experienced place.

Heidegger was not a geographer;[1] nonetheless, his conception of how people, environment, and time are bound together into a unity that provides for the appearance of the world relies on a particular phenomenological approach to knowing. Such an approach requires attention to the ways in which things and places appear and are subjectively experienced, grounded in knowing place with our bodies, emotions, and intellects (Merleau-Ponty, 1945; Paterson, 2009; Tuan, 1979). Experiencing and knowing do not rely on a theory of sense perception with subsequent construction of representations. Rather such perspectives posit people as an integral part of a continually emergent world and not as 'mere' observers (Ingold, 2001); they are embodied. Our muscles, our sense of balance, necessarily respond to places as we move in and through them, and vice versa. Further, this response of and in our bodies is not purely dependent on our physical capacity, and the physical demands and sense offerings of place. Indeed, our bodies carry within them memories, experiences, and habitual – even largely unconscious – social and cultural norms that affect and are affected by our involvement in the world. Our personal and social knowledges are not solely intellectual, but embodied and included in the ways in which we experience places (Casey, 1996; Paterson, 2009).

In this light, the present work sought to base an understanding of place at Cape Bruny on just such an embodied and intertwined approach to perception. A broadly phenomenological method has been used – one that required paying attention to and journaling consciously embodied experiences during two field trips to Bruny Island with fellow geographers. The aim of the first visit in 2013 was a broad study of islandness and island place using autoethnographic and haptic reflections and journaling. The second visit in 2014 considered the particular place that is Cape Bruny. The study that occurred during that visit draws on the recorded impressions, experiences, and reflections of colleagues during a three-hour visit to Cape Bruny on the second field trip, as well as on the memories of a former lighthouse keeper's wife, opportunistically met. A layered analysis that considers these recorded experiences and reflections and their meaning is drawn out in what follows and in light of the foregoing material on place. The chapter is structured to provide a brief view of the geography and history of Cape Bruny and its lighthouse, followed by more detailed description and analysis of the subjective experience of being there. Attention first turns to outline some of the features of the island that is Bruny, and of the lighthouse at Cape Bruny. How the lighthouse is experienced by colleagues and self is then considered, and emphasis placed on the lighthouse keeper's wife. In the latter part of the chapter, commentary turns to the question how is one to make meaning from the experiences of the lighthouse? Then, several conclusions about the salience of these matters to human geographies and island studies are made.

Bruny Island and its lighthouse

Bruny Island is located off the south-east coast of Tasmania, Australia. As Stanley (1991) records, the European history of Bruny Island reflects seventeeth- and eighteenth-century European competition for territory, colonial expansion, and settlement. The island was first recorded in 1642 by Abel Tasman (1603–59), and there were both French and English landings there in the latter part of the eighteenth century, including one in 1773 by Tobias Furneaux (1735–81), who located a sheltered harbour that he named Adventure Bay after his ship. Subsequent visits were made to the Bay by James Cook (1728–79) in 1777 and William Bligh (1754–1817) in 1788, 1782, and 1808. However, while occupied exploring the Australian coastline it was French naval officer Antoine Bruni D'Entrecasteaux (1737–93) who established in 1792 that this place, now his namesake, was indeed an island.

On its eastern side, the island is separated from the Tasmanian mainland by what is now known as the D'Entrecasteaux Channel. The Southern Ocean lies to the south. Approximately 100 kilometres (62 miles) in length from north to south, Bruny Island is really two distinct landmasses connected by a narrow sand spit. North of the isthmus the island is constituted by ancient sediment and somewhat sheltered from the extremes of weather and ocean. However, South Bruny is the product of a Jurassic dolerite intrusion through this sediment, resulting in landscape fringed by dramatic cliffs, the surrounding ocean dotted with smaller, ragged islands, and submerged rock. The southern coast, and Cape Bruny on its southern tip is exposed to the force of the roaring forties and the Southern Ocean (Tasmanian Government Department of Primary Industries, Water and Environment, 2008).

British settlement of the island commenced in 1818 and soon supported whaling operations, farming, sawmilling, stone quarrying, and coal production. As in other parts of Tasmania, European settlement also brought violence, disease, and displacement to the original inhabitants of Bruny Island, the Nuenonne people, who had inhabited it for many thousands of years. Truganini, the woman who in European histories was often incorrectly and – in terms of Tasmanian Indigenous identity – damagingly referred to as the last Tasmanian Aborigine, was a Nuenonne woman of Lunnawannalonna (the Indigenous name for Bruny Island).

Today Bruny Island has a population of circa 771 permanent residents, the majority of whom (550) live on South Bruny. At certain times, this population is swelled by perhaps as many as 1,500 'Shackies' – people who maintain holiday homes or 'shacks' on the island. The main industries on Bruny Island are tourism, agriculture, and aquaculture and, given an ageing population with many retirees, business owners and employees often commute from mainland Tasmania. Access to Bruny Island for residents, shackies, and workers is via vehicular ferry (on which, more generally, see Vannini, 2011a, 2011b). The ferry also carries up 125,000 visitors to Bruny Island each year. Located on the southernmost edge of the island and separated from population centres by a national park, the lighthouse at Cape Bruny is one of the promoted attractions on the tourist route (Ferrier *et al.*, 2011/2014).

The catalyst for the construction, from local stone, of a lighthouse on the cliffs at Cape Bruny was the shipwreck of the British penal transportation convict ship, *George III*, on 12 March 1835, in which 130 of an estimated 300 people died (Stanley, 1991) – almost all of them convicts. In literature and tourist information, the wreck of *George III* is at the core of the 'foundational' story of the construction of the lighthouse to protect human life. However, planning had started ten years earlier for a lighthouse to protect commercial interests and guide merchant shipping into the port at Hobart. Indeed, arguably even the lives lost on the *George III* were of commercial rather more than humanitarian interest to many, being mainly convicts and therefore sources of indentured labour. It is worth noting that the convicts who died on the *George III* were kept below decks at gunpoint as the ship went down; their deaths were not purely for want of warning from a lighthouse. Nonetheless, those deaths and the loss of the *George III* created a sense of urgency and the lighthouse was completed in 1838 using convict labour. Until 1996, the station was staffed by a series of lighthouse keepers and assistants who lived there with their families. Today the Cape Bruny lighthouse and reserve is a heritage area and tourist attraction with a museum, information boards, delineated paths and busy asphalt car park catering to private vehicles and tourist buses.

Experienced place at Cape Bruny lighthouse

Attention now turns to a detailed reading of place at Cape Bruny as experienced, recorded, and reflected upon in field journals and conversations by self and colleagues.[2] Despite differences in their ages, genders, ethnicities, and focus upon human or physical geography, the contributors' journals reveal a commonality in subjective experience at Cape Bruny in stark contrast to the objective reality of the managed tourist destination described above. The following paragraphs in italics are inspired by those journals using a structured and iterative process of immersion, thematisation, and summarisation. This process is an adaptation of a phenomenological research method (Groenewald, 2004), and provides for knowledge and meaning creation from subjective, first-person responses without reducing the affective and personal purely to data sets or abstracted units of analysis. In brief, the approach involves an iterative process of immersion in the journal contents, and enables one to draw out, categorise, and cross-reference emergent themes to identify commonalities and divergence.

The extracts provided here are a distillation of colleagues' experiences and the reflections of the wife of the lighthouse keeper, supplemented with quotes drawn directly from individual journals and with interpolations of observations, contrasts, and comparisons. The order in which these descriptions are presented was structured by emergent themes, and represents a choice by me to create a journey around Cape Bruny in a way suggested by the voices in the journals themselves. The description of the experience at Cape Bruny therefore starts from the land on which we stood and from which this place was first apprehended by us, and then presents the ocean that all but surrounds the land, the weather that enfolds land and sea, the people within that enfolding and, finally, the qualities of the lighthouse itself.

Land

Cape Bruny is beautiful, rugged, raw, and powerful. Wild, untamed, and untouched, this place is an ancient place. The cliffs that thrust out into the sea are Jurassic dolerite, a material marker of the geological upheaval 60 million years ago. However, these rocks and this shore are not mere monuments to ancient history. They tell a story of continuous and continuing change. The apparently solid rock has been eroded and fractured, forming columns, sea cliffs, rock falls, and jagged edges extending out into and under the sea. Deposited sand provides sandy beaches and the edge shifts and changes with the tides, ocean currents, and storms. 'Nothing stays the same here.' This coast speaks of past, present, and future change. It gathers deep geological time into this place and the lighthouse that stands on the cliff top, gathers that deep time into its structure as just one among many changes wrought on the rock at Cape Bruny. The lighthouse partakes of the solidity and, on a human scale, the imperviousness of the rock. As one person observed of the lighthouse, 'it feels ancient' and, in a sense, it is as old the cliffs on which it sits. In another sense it feels fleeting, impermanent in this ancient place, much like the people who built it or stayed with it for a time.

Cape Bruny is not untouched. The lighthouse, buildings, roads, and tourist infrastructure attest to human presence and significance both in the past and present. However, these artefacts also obscure an older presence. Bruny Island, Lunnawannalonna, was the home of the Nuenonne people for millennia and, with British colonisation, it also became a place of death and displacement. Nuenonne stories are not told here although, according to information that seems to be 'cut and paste' across tourism and parks information, material traces of their lives do remain: 'The park contains a number of important Aboriginal sites, mainly in the form of middens, quarries and artefact scatters. There is also a number of stone arrangements along the coastline of the park' (Tasmanian Government Department of Primary Industries, Water and Environment, 2012, n.p.). However, the Nuenonne were absent in both the perceptions and reflections contained in all but one journal of the experience at Cape Bruny, and arguably it is in this absence that the most poignant intimation of the recent history of the original people of Bruny Island can be found.

Sea

Below the lighthouse, waves lash, crash, smash against these jagged cliffs and rocks, echoing the continuous, rhythmic sound of the sea. The ocean behind the waves heaves, rolls, and whispers. One person described the waves as 'powerful, terrifyingly powerful – the hands of the enormous entity that is the ocean, stretching back behind'. However, the ocean takes on a different aspect from the top of the lighthouse where the eye is drawn out to the horizon. The earth below loses its texture, smells, and sounds. The scrubby coastal bush looks like a carpet and the eye is drawn to the empty 'thoughtless void' of the ocean. The ocean from this vantage point is vast and daunting. Eyes seek the horizon and between the top of the lighthouse and the horizon there is nothing, only the expanse of the ocean. There is no land; only sea, only ocean, nothing but emptiness between here and Antarctica. The ocean is muted; it loses its voice from the top of the lighthouse. There is only quiet surface, a place for ships to sail upon. These ships carried explorers, colonists, convicts, traders, and whalers. However, for most, the people on board those ships have little substance and those who sailed after the lighthouse was built are primarily understood as people who, having survived a long sea voyage, received an ambiguous and ambivalent message of warning and greeting from the lighthouse, across the surface of the sea.

The onomatopoeic words used to describe the ocean are words of violence, destruction, and fragmentation and yet, when the perception of the sea is reduced to the visual by distance, the ocean loses its physical and affective immediacy and threat. It becomes an object conceived of as empty except for purposeful human activity. Indeed, people too become abstracted, imagined as essentially transitory and defined by their purpose. As Hay (2013) remarks, such purely visual engagements with the surface of the ocean hide the complexity of its structure and ecology and the effects of human involvement with it; this can and does have negative consequences.

Beyond the paths of ships and trails marked by lighthouses on its surface, the ocean bears other marks of human interaction in the form of pollution, habitat and species modification or loss, and climate change but these are not immediately perceptible to the eye or contained within concepts of the ocean as purely surface and highway or, for that matter, as wild and untamed.

Weather

Spanning the land, ocean, and their shifting meeting places is the weather. Even more than the features of the land or the presence of the ocean, the wind dominates this place. The wind at Cape Bruny is powerful, raw, wild, and terrifying. The wind is personified; it is the savage wolf that lurks in the wild, dark places of fairy tales (Hay, 2002); it howls and bites and drives the ocean crashing against the shore; it lashes your face; it drives the rain against your skin like a thousand tiny cuts of the knife and 'strips paint from the buildings to its bare base'. This is a wind that is written in this place. The vegetation is thick, low, tough, with spiny leaves. It crouches and leans before the wind. The lighthouse too is a response to the weather in its function but also in its very shape, which is a concession to the force of the wind in this place. Conical, the lighthouse does not present a face of futile resistance to the wind but rather sits within it and allows the wind to pass around it and, in times past, through it, in order to keep the lamp burning (see also Ibbotson, 2001).

These descriptions of wind and wild weather are not representative of the day that my colleagues and I were at Cape Bruny, which was unusually calm with a gentle breeze. While geographic clues and knowledge of the area would certainly indicate strong winds and storms, the descriptions given of the weather were not statements of fact nor potentiality of the weather's power but rather emotionally coloured, almost visceral paintings of the experience of being within it. Ingold (2007, p. 30) has written that we perceive weather in the shape of the landscape and that 'every tree in the arc of its trunk and twisting of its branches bears testimony to the currents of wind' and this seems reflected in the perception of the weather at Cape Bruny.

Lives

> *This howling wind blows in and through an isolated and lonely place. Variations on the word 'isolated' – often just the word – echoed through the journals kept of our visit to this place. Cape Bruny is almost surrounded by the sea and separated from the rest of Bruny Island by forest. It is an island on an island that felt cut off and separate. 'I feel as if I am a long way from anywhere' as one person expressed it. The people who lived and worked at the lighthouse were also seen as isolated. Their lives were solitary, 'lonely and desolate', and their isolation was emphasised by the disconnection of seeing ships 'passing but never stopping'. Some people found the idea of such isolation attractive and liberating, while others felt that being cut off in a very small, close community would be claustrophobic. However, isolation was predominantly expressed as a deep sense of loneliness. In the midst of a journal that was largely detached and fact based, one person wrote, 'lonely – it must have felt so lonely'.*

Another perspective on this isolation emerged in conversation with the wife of a former lighthouse keeper who lived with her husband and young children at three Tasmanian lighthouses, including at Cape Bruny. She described the lighthouse as an isolated place, in many ways closed off from the rest of the world. Hay (2006) writes of the resilience and resourcefulness that has been attributed to islanders in connection with such isolation and this was echoed in the woman's description of life at the lighthouse. She said that living in such isolation was difficult but she also asserted that self-sufficiency was the key to a good life. Despite the difficulties such isolation created, for her it was also one of the most positive aspects of life at the lighthouse. It provided for an enfolding and binding together of her family into self-sufficient and interdependent unit; just them, without the world to interfere. Her story is returned to in more detail in due course.

> *This place is 'haunting'. Cape Bruny is peopled with glimpses of those who have lived and worked there. The sense of loneliness and isolation and the violence and hostility of the wind and ocean is felt on behalf of past Lighthouse Keepers and their families. These people from the past inspired sympathy and sorrow for the hardship that they endured, and the graves of two small children speaks of the difficulty and harshness of life for families here. These graves are 'sad' but they also evoke memories of childhood. One person wrote of the fun that they had jumping on the rocks on the*

beach and wondered if the lighthouse children did the same. However, it is the figure of the Lighthouse Keeper who is most present in this place. He is imagined working all hours through violent storms and is admired for his bravery and sense of duty. There is 'awe at [his] strength' and a sense of being 'overwhelmed with respect' for him. The Lighthouse Keeper does not have an 'easy job' but it is important. In addition to courage and strength he must also have wide ranging knowledge and be able to cope with the routine and mundanity of repetitive and regulated work. Two people noted that the lighthouse and the workers were here to protect trade and colonial interests. However, the dominant image of the Lighthouse Keeper that emerged from the participant journals was of a man, at the 'end of the world', heroically saving lives from the dangers of the ocean.

Again, the indigenous people of Bruny are absent in this experience of Cape Bruny, but they are not the only ones. Women, too, are either wholly absent or peripheral in the field journals recorded by my colleagues. Identification was primarily with the male figure of the Lighthouse Keeper, even though just under half of the participants were women themselves. When women were referred to in the journals it was as an undifferentiated part of the 'families' who experience loneliness and hardship by virtue of living here with the lighthouse keeper. On the one occasion that women were specifically mentioned in a journal, they were given little agency. It is suggested their situation was worsened by the fact that, like the convict lighthouse builders (who are also largely absent in this account of Cape Bruny), they had no choice about being here. However, the histories of the lighthouse indicate – albeit briefly – that women took active roles in lighthouse duties alongside their husbands. The first woman to live at the lighthouse also raised 12 children there, and yet there is little indication of the domestic activity that this necessarily entailed. In other accounts, another woman is mentioned; she slid down one of the rocky cliffs to attempt to rescue a fisherman who had fallen – reportedly at the expense of shredded clothes and modesty (Stanley, 1991). The death of the fisherman is noted in tourist information but there is no mention of the woman who tried to save him (Tasmanian Government Department of Primary Industries, Water and Environment, 2008). How many such stories about women of the lighthouse have been elided?

Lighthouse

The lighthouse is embedded in this isolated place of wild nature, and brave and resilient human beings. Stark white against the green of the vegetation and the grey of the sky, the always visible lighthouse is strong, solid, and immoveable. It is 'the only beacon of humanity for kilometres'. Indeed the lighthouse is seen almost as a person. From a distance it looks like one, standing as a 'steadfast and silent watcher and sentry' at the highest point of Cape Bruny. It seems to have great strength as it stands inside nature and

Figure 3.2 Skeleton. Photograph by Thérèse Murray.

defends and protects, but it is weary, fragile like a human, and vulnerable to the elements and the passing of time. Inside, its 'walls bleed rust'. The 'metal skeleton' of the staircase coils up to the 'warm heart and head', the light, encased in the 'cold body' of the lighthouse (Figure 3.2). The skeletal stairs must be climbed to reach the top of the lighthouse. The ascent creates a new sense of the lighthouse's isolation. As space opens up beneath each grated step, the Earth itself feels as if it is being is left behind until, at the top of the lighthouse, the eye can see what the light sees. Land and water are spread out but flattened, in a circle with the lighthouse at its centre. It is a beautiful, awe-inspiring view, but affectively ambivalent; empowering and daunting in equal measure. The lighthouse stands on the edge, it is a last human outpost at the end of the world.

Boundaries are implicit in the foregoing descriptions of ocean, land, and isolation, but in the personification of the lighthouse standing lonely and vigilant at the edge of the world the edge of the island is made explicit, bringing with it the tension between islands as closed, contained, isolated, and simultaneously open and porous with networks extending in all directions, what Baldacchino (2003) calls a nervous duality. My own initial reaction to the lighthouse was that, sitting as it does on the edge, it encompasses some of that duality. It is a marker of the extent of territory, it is explicitly in place to facilitate connections and yet it also warns of danger. It welcomes the traveller but also says 'Danger, keep back, back off, do not come too close!' The lighthouse delineates and makes explicit edges and boundaries, yet blurs them by connecting the sea and land with light. It suggests that edge and boundary are not as clear as they appear to be.

The lighthouse keeper's wife

The lighthouse is understood by the former lighthouse keeper's wife as a far more personal and intimate place than that described by my colleagues. She said that she loved lighthouses. When asked why, she did not give a reason. Instead, in a powerful example of embodied memory, she told and unconsciously acted out a story about using her body as a shield for her family against the weather. Doubled over in her seat, with her arms wrapped around her body, she spoke of enfolding her children in her arms, holding them tightly to her body against the force of the wind as she made her way up a path to another lighthouse on Matsuuyker Island, located south of Tasmania in the Southern Ocean. Once there, she and her children were welcomed into the lighthouse by her husband, isolated from the outside world but enfolded with the warmth and security of both family and lighthouse. This description seemed to encapsulate her sense of life at the lighthouses in which she lived; places that, in their isolation, created a bounded and secure place in which her family could flourish, and provided for their resilience and interdependency.

The identity of the former lighthouse keeper's wife, at least to the extent that she shared with me, is bound up with place at Cape Bruny and with the lighthouse in particular. She seemed to see the lighthouse at Cape Bruny as home, even though she had not lived there for almost 60 years. The lighthouse is the resting place for her husband's ashes; he has gone 'home' and yet he remains present at her current home at Dennes Point on the northern tip of North Bruny Island too. In her sunroom are three scale models that her husband made of 'their' lighthouses; Matsuuyker Island, Cape Sorrell on Tasmania's west coast, and Cape Bruny. In her garden is another lighthouse he made – a scale model of Cape Bruny lighthouse (Figure 3.3). Although the lighthouse at Cape Bruny no longer pierces the night, this small echo of it does. At five o'clock each night, the system that her husband created brings the little light to life again. In her elder years, the former lighthouse keeper's wife no longer resides full-time in that small Dennes Point house, but each night the lighthouse in her garden builds a bridge across space and time and place, and the Cape Bruny lighthouse is made present to her, alongside her husband, tending the light as he did in life. She says that she will 'stop' when the lighthouse does, and although she did not explicitly say so, I suspect she will then join her husband at the lighthouse at Cape Bruny.

The foregoing narrative is intended to give a sense of the experience of place during a visit to Cape Bruny and the lighthouse, and to hint at the changeability of the appearance of place by reference to the contrasting

Figure 3.3 A bridge across space and time. Photograph by Thérèse Murray.

recollections of someone who has a long-lived and deeply personal relation-ship with the lighthouse. The following narrative now seeks to tease out some of the meaning revealed during one particular interaction in an island place, and it leads to a point of conclusion in which I question the relevance and application of such particularity and specificity to an understanding of place and, in particular, islandness and island places.

Making meaning

Like the great giant Christopher it stands
Upon the brink of the tempestuous wave,

Wading far out among the rocks and sands,
The night-o'ertaken mariner to save.
                        (Longfellow, 1849)

'The Lighthouse' by American poet Henry Wadsworth Longfellow is contemporaneous with the building of the lighthouse at Cape Bruny and it depicts a romantic vision of the valiant lone lighthouse, standing firm against the untameable tempest and ocean. These lines, and much of the balance of the poem, use Romantic tropes placing the rugged individual – in this case a personified lighthouse – within and in opposition to the awe-inspiring sublimity of nature. I include this poem as an example of what is a ubiquitous Western representation of historic lighthouses and, by extension, their keepers. This representation incorporates ideas of nature as set apart from and in opposition to human beings; sublime nature as a crucible for rugged individualism (Cronon, 1996; Hay, 2002) and the romance of lighthouses as brave saviours of those at sea (Phillips, 1971). Such representations persist into the twenty-first century in relation to nature and lighthouses. The preface to a book on lighthouses published in 2001 suggests that 'lighthouses provoke a special passion because they remind us of when ships ruled the world', the 'grim perils of the sea' and are the 'first sign of civilisation after a long sea voyage' (Davison, 2001, p. i). Promotional material for another book published in 2013 quotes Longfellow's poem (Seaside Lights, 2013), while the book itself (Searle, 2013) is prefaced on the cover page with a biblical verse, Isaiah 9:2 – 'The people who walked in darkness have seen a great light: they that dwell in the land of the shadow of death, upon them hath the light shined.'

As Fletcher (2011) argues, such persistent representations have a performative role in our constructions of experienced reality. However, as much as a sense of place is mediated by the cultural representations that we carry into a place with us, it is also evoked by the material, physical actuality of being in a place (Hay, 2013). Places, according to Tuan (1979), are ultimately sense bound and, as Casey (1996) argues, it is the particularities of place that produce and invoke cultural understandings of places. My colleagues undoubtedly brought to bear in their experience of Cape Bruny what Fletcher (2011) calls performative geographies of island places including sublime nature in opposition to the human beings. These add to and layer the universalised idea of the romance of lighthouses. Further, as Maddern (2008) explains, the way in which places are experienced by visitors is also framed by the management choices that are made in the presentation, exclusions and storytelling at the site. At Cape Bruny at least some of these decisions clearly perform an heroic and romantic view of the light and nature. This romanticising tendency is reflected in my colleagues' personifications of both the environment and the lighthouse, and in the characterisation of the Lighthouse Keeper. However, in addition to such representations and cultural constructs there was also a felt experience of being at Cape Bruny. The power of the waves, sight, sound, smell, and taste of the ocean, the dominance of the wind and the material evidence of people's lives, including the lighthouse, were all available to the senses. This physicality, the potential danger for earthbound beings, is a ground for such romantic images of lighthouse, as well as being mediated by them.

There is nothing inherently dangerous in the physicality of Cape Bruny. The weather and ocean are powerful but ocean, shore, wind, land, and lives have been intertwined and changing in response to each other for millennia. This place is violent and dangerous only within a particular relationship between human beings and the environment where the particular usage and human purposes are threatened. This relationship was, and in many respects still is, one in which people understand the ocean primarily as a means of transporting people and goods, and as a space over which power is exercised between places (Dening, 2004; Steinberg, 2001). Prior to the lighthouse being built at Cape Bruny, multiple shipwrecks had demonstrated that the physical environment of the coast at Cape Bruny was genuinely dangerous to ships and those on board then, and therefore an obstacle to passage between places (Stanley, 1991). Coupled with the dangers to shipping at this particular place, this understanding of the ocean as highway (Steinberg, 2001) is the reason for the existence of the lighthouse and also therefore, the reason for the presence of European people with their cultural and social constructs in this place at Cape Bruny. The lighthouse is an attempt to manage the contextually understood violence and danger of the ocean through technology (Ibbotson, 2001; Stanley, 1991). As such, the lighthouse sets human ingenuity and civilisation in opposition to those dangers. This opposition, embodied in the lighthouse, foregrounds an understanding and a reality of nature in this place as dangerous to human beings.

The purpose and function of the lighthouse was not to remove such danger but to warn against it. The lighthouse acknowledges the futility of resisting the wind in its shape (Ibbotson, 2001) and the light reaches across the surface of the ocean, but it cannot affect the ocean itself. The lighthouse regulates the movement of ships, not nature. Nature is therefore framed as still wild by the lighthouse. Despite no longer being in service, its presence still performs this function as marker, a reminder. In conjunction with enculturated representations of wild nature (Fletcher, 2011; Hay, 2002), this aspect of the lighthouse contributes to an understanding of Cape Bruny as wild and untouched, despite the undeniable presence of humans and their infrastructure on land. Ingold (2010) claims that human beings leave cumulative traces and paths in the places that they stay and pass through, and the land at Cape Bruny is replete with such traces. The lighthouse also speaks of human traces on the ocean even though, as the participants experienced at the top of the lighthouse, the ocean surface appears as a plane or blank surface where the traces of humanity are not as obvious as on the land. According to Steinberg (2001) modernity constructs this ocean as effectively a non-place, but the lighthouse and its lamp contradict this. The lighthouse confirms the presence of people at sea and, at night, the lamp illuminated the boundaries of safe paths on its surface. Even now, when the light no longer functions, the eye can follow the line of the light through the air and across the water along the paths it once made. It is, to use Maddrell's (2013) term, still present in its absence and in its spanning of the elements and its tracing of human paths it draws together people, land and the still wild nature of the weather and ocean.

The fact that Cape Bruny is bounded on three sides by this ambiguous ocean may contribute to the strong sense of isolation felt by participants. As is the case for the edges of most islands, the sea here is a very real 'perceptual boundary' (Hay, 2013, p. 216). The lighthouse itself provides a strong, visual evocation of isolation. It stands high on a hill, apart from other signs of human presence and both from a distance and from the top of the lighthouse itself, it appears to be largely surrounded by dense vegetation and the sea. The experience of climbing the lighthouse stairs and the sense of leaving the earth behind to stand at the top of the lighthouse may also be an isolating experience. Ingold (2010) calls human beings terrestrial creatures; ascending the lighthouse stairs, with the air beneath our feet emphasised by the gaps in the stairs, took us into a different realm from our familiar and natural place on the ground (Tuan, 1979). Getting to the lighthouse also required the type of dedicated journey over both water and land that, as Royle (2001) argues, contributes to the sense of remoteness and isolation associated with islands. The lighthouse is approached not only over water to Bruny Island, but via a dirt road through bush land that separates it from the rest of South Bruny, and then uphill by foot on a path bordered by scrubby vegetation until the lighthouse is finally reached. The lighthouse at Cape Bruny is located on an island itself surrounded and seemingly isolated by bush and ocean, features that provide apparently delineated edges and boundaries; it feels like an island on that island.

It is therefore perhaps not surprising that performative geographies of island places are apparent in the descriptions of the isolation of Cape Bruny (Baldacchino, 2004; Fletcher, 2011). These performative geographies include the traits of resourcefulness claimed to be characteristics of island peoples (Hay, 2006) and these characteristics were ascribed – not without good basis in material reality as the reflections of the lighthouse keeper's wife attest – to the people of the lighthouse. However, speculation about the lives of lighthouse keepers and their families went beyond a conceptual understanding of probable hardship. Islands are often thought of as closed and disconnected from the outside world (Baldacchino, 2004; Royle, 2001) and the disconnectedness of the lighthouse was not only acknowledged but felt as a strong sense of loneliness. A number of people also saw Cape Bruny as prison-like suggesting that the women and families who accompanied the lighthouse keepers, like the convicts that built the lighthouse, had no choice in being here. This sentiment echoes the representation and the historical and contemporary reality of islands as carceral places, which has a particular resonance in Tasmania with its history as an archipelagic penal colony (McMahon, 2003).

Such sense of entrapment is in stark contrast to the description provided by the woman who had lived there. For her, the lighthouse was her family's island on an island, providing a sense of security and freedom and, as Royle (2001) observes of islands more generally, both a separation and refuge from civilisation. The difference between her story and my colleagues' perceptions of isolation is a clear demonstration of the difference between being 'contained within' rather than 'imprisoned behind' (Hay, 2006, p. 216). The sense of loneliness and imprisonment expressed by my colleagues may, with further investigation, reveal something about the writers' own lives. Tuan (1979) claims that the experience of travelling to unfamiliar places reveals and increases the awareness of our own place and identities, and from that perspective the loneliness and sense of entrapment felt by my colleagues at Cape Bruny may also be an expression of the contrast that it presented to their own urban, mobile, and hyper-connected identities.

The sense of isolation that was described also contains within it the same contradictions and dualities thought to characterise islands, in particular the tension between being open and closed (Baldacchino, 2005; Hay, 2006). The lighthouse was an integral part of a network of trade and power and, despite the expressed sense of isolation, my colleagues did record evidence of Cape Bruny's connection to the wider world. The factory time clock in the lighthouse hints at connections into a management hierarchy that exerted external control and regulation of activities at the lighthouse. The lighthouse mechanisms, glass, and lamps were all transported from England. Fuel and other supplies were brought across from Hobart, many of them originating from locations across the globe (Ibbotson, 2001). The lighthouse keepers also came to work here from other places and often returned to those places or travelled elsewhere after a stint at the lighthouse (Stanley, 1991). Now it is mainly tourists who visit the lighthouse but this is not something new; pleasure cruises to the lighthouse were frequent throughout its history and official visitors also often brought their families and friends with them (Stanley, 1991). The lighthouse at Cape Bruny has existed for over 170 years and, for all of that time, this has been as Hay (2003, p. 23) puts it, a place of 'comings, goings [and] stayings'.

In most of the journals of our visits to Cape Bruny 'the lighthouse keeper' is a single figure who represents a succession of lighthouse keepers that have come and gone at Cape Bruny. This composite figure of the lighthouse keeper incorporates the romantic representations of lighthouse life previously discussed (Fletcher, 2011). However, real lives leave real traces (Ingold, 2010) and there are certainly traces of actual lives at the lighthouse. Maddrell (2013) speaks of bringing the absent into the present through the material memorialisation of the dead, an effect seen in responses to the children's graves at Cape Bruny. Further, as Maddern (2008) writes of Ellis Island in New York, buildings can also speak to us and, in their materiality, preserve and evoke the memory of people of the past. In this sense, the lighthouse itself also recalls the absent. The darkness of the lighthouse lamp bears testimony to the absence both of those who once kept it burning and a mercantile colonial world of seafarers who relied on its signal, bringing them and their purpose into the present. The lighthouse as a workplace, a piece of Victorian industrial culture, speaks in its form and the technologies on display – including a time clock – of the not-so-romantic reality of that work. Although the heroes of the light remain, they are also revealed as men with diverse knowledge and skills who worked within a system of processes, rules and regulations, with a technology that dictated what and when work must be done. However, as Maddern (2008) notes, presences are also evoked from what is absent and not preserved or told in a place. The absence of women's and Nuenonne stories and lives in the experience at Cape Bruny is one such absence that paradoxically highlights a haunting presence.

Maddern (2008) argues that such presences are not merely a cognitive recognition of presence as the opposite to absence, but are a genuine and affective experience, a claim that seems true of the emotional response of my colleagues on behalf of the lighthouse keepers and their families. However, I would argue that presence of people of the lighthouse was evoked through memorialisation or material traces *and* actively invoked by the participants in this study. Lloyd (2013) refers to Smith's (1759) argument that to have sympathy for someone requires an exercise of affective imagination in which we imagine ourselves not *as* another person but in another person's place. For Lloyd and Smith this relation is an ethical one. Reason must be applied to affective imagining, without disregarding the emotional, in order to come to an ethical position about a given situation. The application of reason as an 'impartial observer' to multiple facets of a given situation imagined affectively, provides a means of deciding a nuanced, flexible, and ethical response. The strongly expressed emotional response to perceived hardship and isolation contained in the journals suggests that the conjuring of the presence of the lighthouse keepers and their families may have been a performance of such affective imagining. The situation imagined is necessarily based on participants' own experience of place at Cape Bruny, reflexively informed by imagining what it would feel like to live there themselves. Participants' affective responses to the lighthouse keepers and their families therefore also reveal aspects of their own identities, lives, and worlds which in this case include attitudes to nature and isolation and a valuing of human resilience, responsibility, and the contribution that individuals can make to society.

By way of conclusion

I introduced this chapter by asserting the fundamental importance of theories of place in human geography given that all human interactions, activities, socio-cultural beliefs and structures, as well as the effects and responses to economic and political processes that are important to human well-being and inequity, are experienced and lived in and among places. The case study at Cape Bruny was approached through one of many such theories of place, which posits place as a gathering together and fluid revealing of the intertwining of the human and non-human world with a specific focus on the role of human artefacts in that gathering. The method adopted was haptic and broadly phenomenological in order to reveal the ways one such intertwining as place at Cape Bruny and its lighthouse was experienced.

The preceding description and interpretation of place is just one synthesis of the rich material made available by this method and necessarily represents decisions about inclusion and exclusion, leaving much remaining to be explored via alternate questions and readings. Any meaning posited is therefore provisional and by no means exhaustive, but rather depends on the framing of the research coupled with the particularity of people, time, and the physicality of being some-where. Ingold (2010) describes places where the multiple paths of human lives become entwined as evolving knots and this exposition of place at Cape Bruny might be considered to be just one small thread of such a knot. Even in such a small thread, place is revealed, in both what is present and absent, not as a space containing objects and people that can be concretely categorised, but as a complex continual becoming or revealing that raises as many questions and avenues for further enquiry as it provides answers. Nonetheless, the description of place provided captures how Cape Bruny was experienced in this instance, furnishing an insight into the multi-vocal and dynamic phenomena of place.

The phenomena that comprise place at Cape Bruny clearly included multiple and varied attitudes, representations, and structures of culture and society, both in place and in the accumulated experience that each individual brought with them, and this may argue for a socially constructed model of place. However, I suggest that while these constructions are clearly an important part of the constitution of place, they are not the whole. Although such received understandings certainly mediate our experience of places (Fletcher, 2011) those places must have an existence to be mediated. Conscious and unconscious social constructs provide the boundaries within which things appear, but only in the context of what is available to be perceived (Tuan, 1979). As Hay (2006, p. 33) argues, meanings arise 'from a perpetual dialogue between the physicality of place and the interactions of people with it' and, at Cape Bruny, there is an accumulated and evolving dialogue between the physical environment and human interaction with its attendant and changing social constructions. Evidence of this dialogue is bound up and persistent in the materiality of Cape Bruny. Past and present social constructs are embedded in the physical changes both subtle and overt that are wrought over time by human responses to the particular physicality of this place and, as Casey (1996) argues, this physicality is also implicated in the creation and evolution of human meanings and responses. The interconnectedness of sea, weather, and land is written in this place, but so too is the ongoing presence of human beings – not as separate or independent producers but as intrinsic and interdependent parts of a dynamic creation and revealing of place.

The lighthouse at Cape Bruny can be seen as a synecdoche for the inter-dependency and complexity of the revealing of place at this site. It is a built and maintained expression of attitudes to the ocean, nature, and engagement with the physical realities of this place in the context of those attitudes past and present. The lighthouse stands in for a range of social structures and meanings, and embodies in both its presence and erasures an eco-political history of territorial acquisition, global and local power structures, and cur-rent values and representations of history. Its physical isolation, alone on the hill, gathers in the loneliness of this place but, at the same time, contradicts it with its function of facilitation of human connection and movement. Like the people who have lived or passed through here, the lighthouse also responds to its physical environment. The lighthouse's height, shape, and position are dictated by human functional requirements and by the particularities of its environment – the height of the cliffs, the strength of the wind, the places of hazard in the ocean, which dictate where its light must reach in order to trace out pathways on the sea. Framed by the boundaries of our own life worlds, the metaphorical language of the lighthouse, visual and experiential, invokes both conscious and unconscious understandings of the whole, the place, in which it stands. However, I would argue that this is more than a metaphor. Synecdoche, of course, is from the Greek *synekdokhe* which means *a receiving together* (Harper, n.d. b) and, like Heidegger's bridge, the lighthouse gath-ers, receives together, earth, ocean, wind, and human presence in a way that accords with the thing that it is, providing for the uniqueness and particular-ity of an appearance of place at Cape Bruny.

It is this recognition and exposure of the intrinsic uniqueness and particular-ity of places that renders most useful a phenomenological approach to place as an unfixed revealing and gathering together that aids our understanding of human beings and world in geographical terms. As can be seen in the exposition of place at Cape Bruny, such an approach does not exclude consideration of economic and political structures and processes that are integral to the under-standing of the production of both well-being and inequity; such structures are intrinsic to place. Nor is this approach necessarily in conflict with functional and process-oriented geographical analysis. Rather, by placing emphasis on the experiential, this method has the potential to ground, complement, and inform such analyses by resisting a tendency to abstract and homogenise lived experi-ence. The effects of human structures both concrete and conceptual all occur *in* places, and are implicated in real lives and communities in those places, with all of the complexity and particularity that our very small thread of place at Cape Bruny suggests. Further, a phenomenological approach to place can reveal the material effects of human processes and structures in specific places and, importantly, can also provide insight into less easily quantified and articulated affective and embodied aspects of the lived experience of people and communi-ties, bound up in values, identity, memory, and sense of attachment to place.

Focus on the particularity of specific places can thus disrupt the imposi-tion of theoretical generalisations and cultural representations onto specific people and places, and this may have a particular salience for the understanding of island places. Cultural and theoretical representations of islandness are pervasive in the Western imagination and, although debated in island studies, are widely used both as metaphors and applied to real islands: isolation, peripherality, closed-ness, places of imprisonment, escape, refuge and self-realisation, and utopias and dystopias (Baldacchino, 2005; DeLoughrey, 2001; Hay, 2006; Royle, 2001). As Gillis (2004, p. 1) writes, 'western culture not only thinks about islands, but thinks *with* them'. Although no doubt such representations of islands at least partly arise out of the experience of the physicality of islands, such representations are neither neutral nor innocent. They construct a sameness of islands and islanders and, simultaneously, an 'otherness' that sets island people as different and apart. But islands are not metaphors and nor are they all the same. Island places are revealed in the unique ways in which the specifics of their materiality has been and continues to be intertwined with human action and interaction across time and space. They are unique places in their own right, places that 'enfold and are enfolded by other [unique] places' (Malpas, 2013, p. 11).

Of course, the assertion of the particularity and uniqueness of place applies to all places but it may have a special salience for islands as physically discrete and bounded places. There is a noticeable congruence between islands, or at least smaller islands, with all of their complexity and ambiguity, and the ontology of place that forms the basis of enquiry in this chapter, which suggests that island locations could provide fertile sites for thinking with such a model and the application of similar subjective inductive methods of enquiry. A phenomenological approach to place as an ongoing gathering and revealing can complement and supplement process and structural focused geographical enquiry. So too approaching island places as unique gatherings that can be revealed in embodied and subjective islander experience can give voice to islanders in the understanding of their places and enhance our understanding of specific island places and of the multiple ways in which 'islandness' manifests. Such an approach may be especially valuable as a means of disrupting cultural representations of islands and geographical imaginaries of 'islandness' in order to understand islands on their own terms given that, as Gillis (2004, p. 4) claims, islands 'have occupied such a central place in the Western imagination ... As master symbols and metaphors for powerful mainland cultures, [that] their own realities and consciousness have been more obscured than illuminated.'

Notes

1 Heidegger's philosophy is inherently spatial nonetheless. His focus on the relation between human beings and the physical world is clearly aligned to the core concerns of human geography and his theories have been influential in debates about spatiality, in humanist, feminist, cultural, and embodied geographies, representational and performativity theories, and the conceptualisation of place. Unfolding from being-in-the-word rather than a 'God's eye' view, Heidegger's attempt to

collapse the dichotomy between subject and object, and between perceived and perceiver, provides a model of inherently reciprocal mutual and continual emergence of world, human identity and understanding. His commitment to such understanding also provides a counterpoint to the reductive abstraction of spatial science, and to the structural or functional perspectives of process-oriented geographies.

2 The principal participants in this research were 31 third-year students at the University of Tasmania; 15 women and 16 men ranging in age from 21 to 56 years. Although most of these students were Australian, a number of international students also participated. Prior to visiting Cape Bruny an exercise in haptic geography was conducted to demonstrate and provide practice in the mindful and embodied experiencing of a local environment (on which, see Paterson, 2009). This practice was then applied the field at Cape Bruny. Contributors to the research kept a field journal of their experiences during an afternoon spent at Cape Bruny and also reflected on their experience in writing and discussion after the visit. The journals and post-visit reflections provided the content on which this chapter is based.

4 Too much sail for a small craft?

Donor requirements, scale, and capacity discourses in Kiribati

Annika Dean, Donna Green, and Patrick D. Nunn

Introduction

[Donors] must acknowledge the fact that in very small communities and administrations, there is simply no official capacity to respond to all those questions and studies and requirements. And I guess that problem will continue to be with us for some time ... if the people who are making the decisions about how to administer the money, ... just come down and have a look and visit the ground, I'm sure that would have an impact on their attitudes, and I'm sure they could devise other ways to meet their requirements and at the same time to provide funds in a much more effective fashion.

(Sir Ieremia Tabai, First President of Kiribati,
personal communication, 29 July 2013)

International development agencies have long been critiqued for their propensity to support top-down, short-lived, and one-off projects that fail to produce sustained benefits for their intended beneficiaries. In response to this tendency, it has become commonplace in relation to development projects to think it possible to improve the capacities of local project recipients to sustain project outcomes. In the Pacific, where small island states are commonly regarded by donor agencies as facing capacity constraints, community capacity-building is ubiquitous, at once referring to everything and nothing, and leading to contestation over meaning. Certainly, analysis of project documents and public relations materials of aid agencies operating in the Pacific region illustrates the common use of this phrase 'capacity' as a descriptor both for what is 'wrong' and 'what is needed' in island countries.

In general terms, a capacity can be social, political, cultural, representational, organisational, intellectual, technical, or financial. In the development lexicon, capacity is often used as a synonym for organisational or institutional development or a 'serious-sounding' way to describe training workshops (Eade, 2007, p. 632). Capacity is understood here as the ability to receive and retain or produce something useful to address current and future development needs. Whether or not an entity – such as a person, an organisation, or even an entire nation – is perceived as having or lacking capacity depends on what that entity is expected to be able to produce, retain, or receive. In this chapter, we suggest that donor agencies operating in the Pacific have constructed discourses of capacity in terms that mask the subjective and power-laden nature of the concept (the question of capacity to do what and according to whose agenda), while ignoring their own roles in producing 'lack of capacity' by imposing conditions and requirements derived from continental experience.

Using a case study focused upon a national adaptation project in the central equatorial Pacific Republic of Kiribati, we suggest that donors' failure to understand islands on their own terms has produced a range of detrimental impacts (Nunn, 2004). Donors have imposed procedures, policies, requirements, and techniques based on continental thinking, perpetuating both real and perceived lack of capacity, and hindering development projects from reaching their own articulated objectives.

Given this context, and by exploring geographies of development in relation to notions of scale, islandness, and capacity, we consider three arguments. First, we suggest that the innate characteristics of island states are not sufficiently recognised and understood in development discourse, policy, and practice. Where island characteristics are recognised, they are often depicted in pejorative and even derogatory terms, implying lesser importance relative to continents. Second, we suggest that instead of understanding islands on their own terms, donors frequently construct discourses of capacity in terms of technical and socio-cultural deficiencies able to be remedied. Finally, we suggest that striving to understand islands on their own terms is critical for moving towards geographies of development and adaptation that are suitable in island contexts.

These arguments are developed by reference to empirical data collected during four months of fieldwork conducted between 2012 and 2014, predominantly on Kiribati's capital island of South Tarawa. Information and insights were sourced from 60 interviews and several focus groups conducted with select people from Kiribati (the I-Kiribati), who are engaged in governance and in administering climate change adaptation project funding, design, and implementation. In addition to interviews and focus groups, participant observation was undertaken during several donor–government workshops and meetings. In addition, the chapter also reports on insights gained from analysing the Kiribati Adaptation Project (KAP). KAP was the first national-scale adaptation project led by the World Bank, and was the first World Bank project in Kiribati. Starting in 2003, it has been rolled out in three phases, and is meant to conclude in 2016.

Kiribati: place and people

The Republic of Kiribati comprises 32 coral atolls and one high limestone island, Banaba, and straddles the equator in the central equatorial Pacific

Ocean (Figure 4.1). The atolls are divided into the Gilbert, Phoenix, and Line Islands. A separate Exclusive Economic Zone (EEZ) exists around each of the island groups, bringing the total EEZ to roughly 3.5 million square kilometres (1.35 million square miles), compared to a total land area of just over 800 square kilometres (308 square miles) (MELAD, 1999). The archipelago is highly dispersed, with 4,000 kilometres (2,485 miles) of ocean between the most westerly and the most easterly island. The capital, Tarawa, is situated in the Gilbert Group (1.4° N, 173.0° E).

The population of Kiribati is 103,000 people; it has an annual average population growth rate of 2.2 per cent, and almost half of the population lives on South Tarawa, resulting in a population density of 3,184 people per square kilometre (0.62 people per square mile) (KNSO, 2012). There, overcrowding exacerbates rapid urbanisation, strains public infrastructure and services, and endangers fragile atoll environments (Storey and Hunter, 2010). The results include biodiversity loss, degradation of coastal and marine resources, and threats to the subsurface freshwater lens (Government of Kiribati, 2012). Poor water quality and sanitation contribute to infant mortality rates among the highest in the region (White *et al.*, 2007).

Compounding these immediate development challenges, Kiribati is highly vulnerable to the impacts of climate change. With the exception of Banaba, islands in Kiribati generally rise no more than two to three metres (six to nine feet) above mean sea level. Sea level rise poses threats in terms of coastal erosion, inundation and salinisation of groundwater, the main source of freshwater for the populace (Falkland and White, 2009). Other significant projected future impacts include increased precipitation, increased air and sea-surface temperatures, and ocean acidification (BoM and CSIRO, 2011). These impacts threaten agricultural productivity and coral reefs, which are critical fish breeding habitat. Climate change impacts on atoll nations are likely to be so extreme that they threaten habitability (see Barnett, 2005; Barnett and Adger, 2003; Connell and Lea, 1992; Hay *et al.*, 2003; Nunn, 2013).

Discourses of smallness, islandness, isolation, and capacity

Anglo-American and continental scholarship has long depicted development in island microstates as problematic – facing particular inherent geographic, demographic, or economic disadvantages and vulnerabilities as a result of smallness, insularity, remoteness, and exposure to natural disasters (Briguglio, 1995; Crowards, 2002; Pelling and Uitto, 2001). Disadvantages seen to derive from those characteristics include (i) narrow resource bases, (ii) excessive dependence on international trade with consequent vulnerability to the vagaries of global markets, (iii) potentially high population densities leading to overuse and depletion of limited resources, (iv) small watersheds and limited freshwater, (v) disproportionate costs of public administration and infrastructure, including communication and transport, and (vi) low export competitiveness due to a combination of high freight costs and limited

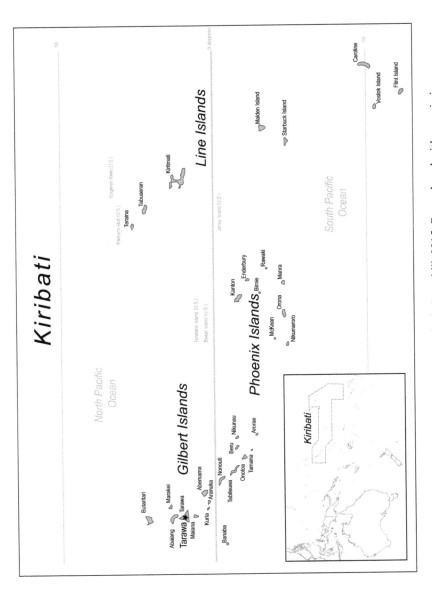

Figure 4.1 The Republic of Kiribati. Source: Nicola Bowskill, 2015. Reproduced with permission.

export commodities. Small populations (in absolute terms) also limit labour specialisation and economies of scale, raising production costs (UNEP, 1994). The economies of island microstates are frequently depicted as non-viable and irredeemably dependent.

This notion that island microstates are inherently vulnerable as a result of their smallness and islandness[1] has been contested by other scholars. Baldacchino (2006b, p. 46) suggests that the absence of 'a large interior, a local hinterland or a local terrestrial core area' in island microstates simply drives them to draw on an extra-territorial 'hinterland' for a range of resources including labour markets, higher education, and investment opportunities. Islands display 'a generous degree of openness and integration with the outside world in order to survive' and are likely to have footprints that extend far beyond their territorial borders (Baldacchino, 2006b, p. 47). The MIRAB syndrome, characterised by (Mi)gration, (R)emittances, (A)id and (B)ureaucracy, has become a common way to describe this phenomenon, although other models and variations have been developed over the years (Bertram and Watters, 1985, 1986).[2] Remittances (from migration) and aid-funded bureaucracies are not supplementary to the economies of island microstates, but are seen to have determined their evolution. That remittances and aid are drawn from outside the national borders of island states does not necessarily produce vulnerability or dependence, so long as such resources and relationships can be sustained.

All nations depend to some degree on other nations for trade and access to commodities not produced domestically. The main difference for MIRAB economies is that foreign exchange is earned predominantly by exporting labour rather than commodities; through aid receipts (which can also be creatively interpreted as a form of exchange, albeit not necessarily fair exchange);[3] and, in the case of Kiribati, interest on the Government's Revenue Equalisation Reserve Fund (RERF)[4] (Bertram, 1999). Given this context, it might be more apt to reconceptualise small island states not as dependent, insular, or vulnerable units, but as open and resilient, sometimes able to draw on a range of resources outside their territorial boundaries and to maintain agency over their economic development.

Hau'ofa (1993) also contests the notion of smallness, making a philosophical point about how imposed discourses of smallness affect people's perception of themselves. He suggests that the portrayal of islands as inherently deficient and dependent as a result of their smallness is the latest iteration of belittling and derogatory discourses about islanders extending back to first contact with Europeans. Early missionary depictions of islanders as 'savage, lascivious and barbaric' have had enduring impacts on how islanders view their own customs and traditions, to the extent that some islanders split their history into two distinct phases: the dark pre-Christian era and the light era following the introduction of Christianity (Hau'ofa, 1993, p. 3). Likewise, discourses of tropicality, associated with the development of tropical medicine in the eighteenth century, portrayed tropical environments as dangerous and unsuited to Europeans, as places where 'disease, decay and putrefaction ran rampant in the moist warm air of the tropics' (Bankoff, 2001, p. 21). Discourses of tropicality further reinforced colonial representations of islanders as uncivilised, backward, and racially inferior (Barnett and Campbell, 2010).

Hau'ofa (1993, p. 7) draws parallels with early depictions of islanders, and argues that smallness is as much a colonial construct and 'state of mind' reified by Western cartographers and geographers, as a reflection of reality. The concept of smallness depends on what is included and excluded. If only land area is included, islands may indeed be perceived as small. But the myths, legends, and cosmologies of the people of Oceania traditionally extended as far as the oceans could be traversed and as deep as the oceans could be exploited, and encompassed the heavens and the stars, home of the deities and key instrument of navigation. Islanders saw the sea as a source of connection, sustenance, and possibility:

> There is a gulf of difference between viewing the Pacific as 'islands in a far sea' and as 'a sea of islands'. The first emphasises dry surfaces in a vast ocean far from the centres of power. When you focus this way you stress the smallness and remoteness of the islands. The second is a more holistic perspective in which things are seen in the totality of their relationships.
>
> (Hau'ofa, 1993, p. 7)

Hau'ofa's focus is upon unsettling how discourses of smallness are perceived and construed, and it is noteworthy that the implementation of the United Nations Convention on the Law of the Sea (UNCLOS) in 1994 defines island microstates as anything but small. The UNCLOS granted nations exclusive usage rights over the natural resources extending 200 nautical miles from their declared coastal baseline. For Kiribati, the vast dispersion of the archipelago poses logistical and governance constraints, but has the beneficial flip-side of enabling claim over an EEZ more than 4,000 times the land area of the nation, vastly increasing resources and economic possibilities. In a similar vein, and in response to the negativism of much recent work highlighting the vulnerability of island societies, there has been an upsurge of interest in the traditional resilience of island societies and their ability – frequently undervalued in development discourse – to weather environmental changes (see Allen, 2015; McMillen *et al.*, 2014).

There are different and contradictory ways to conceptualise islands: as small, closed, isolated, and vulnerable; or as integrated, mobile, and resourceful. Islanders' lives are influenced by this dialectic between openness and closure – affected by discourses of insularity and isolation on the one hand, and by migration and mobility on the other (Baldacchino, 2004; Connell, 2013). While continentalists have portrayed islands as mere appendages to continents, a dedicated strand of scholarship – termed nissology – has endeavoured to understand islands 'on their own terms' (McCall, 1994, p. 63; also Nunn 2007). Nissology is a 'science of island thinking' that endeavours to 'turn the dominant continental paradigm on its head', emphasising the agency and strengths of islands and islandness, and the relative nature of constructs such as smallness, insularity, and isolation (Depraetere, 2008, p. 17). In relation to perceived capacity constraints, a pertinent question then is: capacity to deal with what and according to whose agenda? Who gets the power to determine what is a sufficient level of capacity, and to diagnose who or what falls short of these expectations and why? Adopting a nissological approach entails both attempting to understand islands on their own terms and accepting islands as they are. The next step entails – for the outsider – critical reflection on one's own assumptions and modes of operation. Problematising islands as 'lesser' versions of continents dangerously implies that building their capacity to conform to continental norms and benchmarks – that is, changing islands to be more like continents – should be both possible and desirable.

In advancing this critique, we do not wish unequivocally and uncritically to denigrate efforts to build capacity. For instance, we believe that for the I-Kiribati training and acquisition of qualifications – particularly certain technical skills – could go a long way to improving self-sufficiency and ownership over development projects and processes. Over-reliance on foreign technical consultants is costly and inefficient, and prevents I-Kiribati from benefitting from, for example, employment associated with such projects and processes. We do appreciate the original intent underpinning community capacity-building approaches, which are intellectually rooted in the liberation theology of Paulo Freire (2005) and Amartya Sen's (1985) work on entitlements and capabilities. Yet, capacity-building approaches are also disabling because, *ipso facto*, they evoke the idea that underdevelopment is the result of deficiencies in the capacities of recipients of initiatives. Such notions have condescending overtones, and they risk damage to personal and perhaps to national self-image (echoing Hau'ofa's sentiments, discussed above). As well as being Eurocentric, capacity-building initiatives tend also to be donor-centred (Eade, 2007). Too often they focus on improving people's abilities to comply with donor requirements, rather than recognising and building upon locally valued capacities. That is, capacity-building initiatives tend to emphasise development of skills such as monitoring, reporting, and donor-stipulated formats for procurement and financial management. Capacity-building initiatives targeted at institutional strengthening often emphasise Western, top-down techniques, such as the framing of legislation and the creation of new government policies. While policies and legislation can be effective, gaps in promulgation and enforcement are common in the Pacific, rendering this approach often ineffective (Nunn, 2010; see also Turnbull, 2004). Drawing on traditional governance systems at the community level is likely to be equally or more successful at giving effect to social change (Nunn *et al.*, 2013).

Adopting a nissological framework may encourage people from outside island countries, including development and adaptation professionals, to reflect on the suitability of their own expectations and the requirements of the institutions which they represent. It may also help them to better understand how these requirements affect their own objectives and – at a deeper level – the material and socio-cultural fabric of islands. Donor representatives might consider how to build their own capacities so that they better align with the cultural mores and expectations of the island people they seek to assist. Donor capacity-building might be achieved by language training and cultural immersion, strategies and practices that are more substantial than those common in many Pacific Island countries for several decades during a period when climate change imperatives have been communicated almost exclusively in foreign languages and by reference to inappropriate cultural contexts. For example, using terms such as anticipatory adaptation or no-regrets adaptation 'may alienate Pacific Islanders, whereas, in fact, they have been doing all these things for decades, even centuries' (Nunn, 2009, p. 216).

Skills, capacities, and donor requirements

Increasing donor alignment with recipient country systems of governance and increasing the 'ownership' felt by recipient country populations are two of the five principles of the Paris Declaration on Aid Effectiveness, which was designed to help tackle the issues that have been hampering the effectiveness of international development aid for decades (OECD, 2005). Yet in the quest for aid effectiveness the host of new requirements for accountability, transparency, and 'recipient country ownership' is placing new burdens on already-stretched Pacific island administrations. This turn of events is due to the disconnect between what is demanded by donors in terms of skills and time to meet donor requirements and the capacity restrictions of small populations.

Small population size inherently constrains the scope and depth of skills available in island microstates. Given that there are limited opportunities for full utilisation of technical skills, these are especially in short supply; this makes it difficult for island microstates to access the expertise they require without seeking external assistance (Haque *et al.*, 2015). Limited specialist technical proficiencies among government officials reduce the influence that island microstates can have in and upon international organisations, exacerbating skills shortages at home while such officials are out-of-country (Corbett and Connell, 2015). Skills shortages are further compounded when those with formal qualifications are lured from the public service to work overseas, for instance in regional organisations or with locally based donor agencies that offer higher remuneration. Although skilled migration is often portrayed as a 'brain drain' problem by donor agencies and others, it is not necessarily seen as an issue for islanders; rather, it is viewed as part-and-parcel of island life, functioning as an 'overflow vent' to stabilise populations at sustainable levels and provide an important source of foreign exchange and household cash flow (McCall, 1994, p. 5). Skilled migration is also integral to the Kiribati Government's long-term climate change adaptation policy to enable citizens to migrate with dignity instead of becoming climate change refugees (Government of Kiribati, 2015). This strategy has the additional benefit of ameliorating both overcrowding in Kiribati due to high population growth *and* migration opportunities, which typically are more limited than those available to Polynesian and Micronesian neighbours – the first of which have strong migration links with New Zealand, the second with the United States.

Haque *et al.* (2015) show that the population penalty faced in small island countries is sufficient to explain underperformance in Public Expenditure and Financial Accountability (PEFA) assessments. That is, countries with smaller populations consistently receive lower PEFA scores, because their governments do not have the capacity to adopt best practice public finance management systems. PEFA assessments measure government performance in a range of areas such as policy-based budgeting and predictability and control in budget execution; credibility via accounting, recording, and reporting; and external scrutiny and auditing. The population penalty means there is a limited pool of practitioners with the specialised skills critical to achieve a sound PEFA rating. For instance, an assessment carried out in Kiribati in 2012 found that nine director-level positions in the Audit Office of the Ministry of Finance and Economic Development remained unfilled on a long-term basis because of a lack of appropriately qualified and available people (Haque *et al.*, 2015).

Achieving a sound PEFA rating is perceived as prerequisite for promoting donor alignment with country systems. For instance, donors require a sound score before they are willing to use recipient country systems for planning and budgeting to deliver aid funds. Use of recipient country systems by donors is seen to promote aid effectiveness by reducing transaction costs, making better use of limited capacity, and increasing recipients' control and ownership over their development. Ironically, if recipients receive a poor PEFA score, donors veer away from use of national systems towards project-based approaches that impose even higher transaction costs and further overstretch capacity. The delivery of international development aid in the form of projects results in duplication and wasted resources. For example, White (2010) points to nine large-scale studies that have been commissioned by international donor agencies on water supply challenges in Kiribati since the early 1990s, each reaching the same broad conclusions (see Donner and Webber, 2014). Project-based approaches also result in weaker national systems of accountability, as the attention of government officials is diverted away from national (horizontal and downward) accountability systems to diverse upward accountability procedures of multiple donors, each according to different timelines. Government personnel often struggle to knit together multiple projects with various objectives that form part of a coherent program of activities capable of progressing national development goals.

Calls for increased recipient country ownership over development in the aid effectiveness agenda has, in practice, 'increased demands on recipients with new conditions over management of aid funds, the setting of development strategies and the meeting of other global obligations' (Murray and Overton, 2001, p. 272). Coupled with the small scale of government in island microstates, new requirements for accountability, engagement, and consultation are perversely and ironically resulting in loss of ownership over development, or what has been termed an 'inverse sovereignty effect' (ibid., p. 272). In the neighbouring country of Tuvalu (8.5° S, 179.1° E), the constant stream of donor-initiated meetings, workshops, consultations, and training sessions, all designed to increase ownership over development, substantially limits the pool of skilled government officials available to carry out core government tasks, and places an enormous burden on the public service. Calculating the number of development-related visitors to Tuvalu every year at approximately 900 people, and assuming each visitor conducts on average six hours of meetings with government officials over the course of their stay, the equivalent of five full time staff – 5 per cent of the civil service – is employed solely to attend these meetings on an ongoing basis (Wrighton and Overton, 2012, p. 248).

An interesting counter-interpretation of the situation described above is posed by Goldsmith (2015). Drawing on the MIRAB thesis, Goldsmith suggests that instead of being a distraction from core business, meeting and greeting development professionals *is* core business for Tuvaluan officials, given that the success of MIRAB economies relies on close political relationships with sponsor states. Moreover, he suggests that, in the case of Tuvalu, the government has deliberately promoted the nation as an icon of vulnerability in order to obtain increased aid rents and revenue by promoting so-called 'dark tourism', which is targeted at development professionals and journalists among others. Webber (2013) similarly suggests that government officials are compelled to enact vulnerability in Kiribati, although she reaches a conclusion different from Goldsmith's insofar as the outcomes of such enactments are concerned. Drawing on notions of performativity, she proposes that vulnerability in Kiribati is *produced*, materially and discursively, in encounters between I-Kiribati bureaucrats, consultants, and financiers; encounters which are 'saturated with power' (Webber, 2013, p. 2722). We question whether the effort and capacity invested by I-Kiribati bureaucrats in such 'performance' is sometimes counter-productive; that is, more than the funding is worth in terms of its ability to alleviate vulnerability. We pose this question knowing that donor consultations with both government and communities are often used as a tool to lend legitimacy to, rather than meaningfully influence, project designs (see Cooke and Kothari, 2001). Can consultations and meetings that absorb valuable time and capacity, but that do not genuinely affect donor behaviour and project implementation, realistically be considered part of the useful 'core business' of government officials? Based on our Kiribati Adaptation Project case study we suggest that the scope and magnitude of donor requirements and conditions, relative to available capacity, can prevent projects from achieving even their intended objectives, and can negatively impact upon sense of empowerment, self-esteem, and cultural identity, thus undermining the goals of both donors and recipients. We also test whether and how participation and consultation can be used to mask power imbalances and lend legitimacy to donor agendas at the expense of local priorities.

The Kiribati Adaptation Project (KAP)

The Kiribati Adaptation Project was the first attempt by the World Bank to assist the people of an atoll country to adapt to climate change, and the first such involvement by it in Kiribati. The project was designed to roll out over three phases: Preparation (2003–5), Pilot Implementation (2006–11), and Expansion (2012–16).

The Preparation Phase (KAP I) was small in scope, employing only a handful of consultants. Outcomes were to include the initiation of a national consultation process to develop project priorities; identify pilot sites for adaptation; and start a process to mainstream adaptation into national planning. The Pilot Implementation Phase (KAP II) began in 2006, and was much larger in scope. Its objectives were to build the capacity of the Government of Kiribati to 'reduce the detrimental impacts of climate change on the fragile atoll ecosystems of Kiribati', and to mainstream climate risk information into national economic and operational planning (World Bank, 2011, p. 2). This work involved trialling both hard and soft adaptation measures focused on coastal protection and freshwater management (World Bank, 2011). Initially envisaged as a three-year project, the deadline was later extended to 2011, constituting a five-year project. The Expansion Phase (KAP III) started in 2012, and remained focused on freshwater management and coastal protection, and on extending activities conducted under KAP II. These activities have included installing rainwater harvesting systems and infiltration galleries, detecting leaks in the reticulated water supply, constructing seawalls, and planting mangrove seedlings. As KAP III is ongoing, discussion below is focused primarily on KAP II.

KAP II: over-ambitious objectives, over-complicated design

KAP II was funded by a grant obtained from the Global Environment Facility (GEF), with co-finance provided by the governments of Australia, New Zealand, and Kiribati. Funds were accessed from the GEF via the World Bank, the official multilateral implementing entity (MIE) for the project. Implementation was executed via line ministries in the Government of Kiribati, with direction from a Project Management Unit (hereafter, KAP PMU or PMU) established for that purpose and oversight from the Office of the President.

Our research indicates that the project design was over-ambitious, over-complicated, and fragmented, resulting from a combination of factors, including project funders' eligibility requirements, World Bank influence over the project design, and the Government of Kiribati's passive stance in the early phases of the project. As part of the eligibility requirements of the GEF's Strategic Priority on Adaptation (GEF-SPA), it required that projects generate 'global environmental benefits' and therefore compelled project managers to adopt an ecosystem-based approach to adaptation.[5] According to the World Bank's own Implementation Completion Report, this require-ment 'likely "forced" a selection of Project activities [to] accommodate both country and donor priorities', contributing to a fragmented project design that spanned multiple ministries and sectors in order to satisfy GEF objec-tives (World Bank, 2011, p. 9).

The project objectives covered five areas: planning and policy; integrated coastal management; improved freshwater supply; capacity-building and awareness-raising among communities; and project management. Multiple activities were designed under each of these themes, with implementation spanning several sectors and ministries, including Internal Affairs (MIA), Public Works and Utilities (MPWU), Environment, Lands and Agricultural Development (MELAD), Finance and Economic Development (MFED), and Fisheries and Mineral Resource Development (MFMRD).

In addition to those requirements, staff in the newly established KAP PMU found themselves having to comply with complicated World Bank policies and procedures, even though the World Bank provided neither funding to the project nor involved itself significantly in its implementation. The KAP PMU team had no experience with World Bank policies and procedures of fiduci-ary management, procurement, and reporting. World Bank procedures were seen to pose an enormous barrier to implementation, as described below by a government official from the Office of the President:

> I think that because it's a World Bank thing, we find ourselves having to follow the World Bank guidelines and conditions for almost every single activity and planning that we need to be doing, with regard to the actual activities that have to be undertaken on the ground. So, it is very cumber-some and limiting on our part, because we virtually don't have the capac-ity and the systems in place to meet the requirements of the World Bank.
> (Interview with government official, Office of the
> President, South Tarawa, 16 August 2013)

The project design required the KAP PMU team to manage more than 70 individual contracts, two-thirds of which were international (Hughes, 2011). PMU staff perceived the World Bank's procurement process as particularly complicated and time-consuming, as articulated by one staff member below:

The World Bank has particular processes related to procurement. First, we need to develop the Terms of Reference for individual consultants. The World Bank has to review that and provide the 'no objection approval' to proceed. So, if it is approved, then we can proceed with the advertisements for those positions … then maybe in a month's time we receive Expressions of Interest for the position. Then we do the evaluation of those candidates, and we have to come up with the Consultants' Evaluation Report, again to seek the World Bank's review and approval for that selection. So we send the Consultants' Evaluation Report to the World Bank to review, then if they come back with the no objection approval – the second approval – we have endorsement for that selected candidate. So then the next stage is the contract negotiation. When we have an agreed draft we sign it with the consultants, and then we send that again to the World Bank for review and no objection approval – the third approval. And in a way, all this takes a lot of time.

(Interview with KAP PMU staff member,
South Tarawa, 4 September 2013)

The World Bank did not have a permanent in-country presence. Distance slowed implementation, with approvals and other communications undertaken by telephone and Internet, the latter being slow and irregular in Tarawa. According to one government official the overall process was too widespread and wide-ranging, and too difficult in a country the size of Kiribati. In large part this outcome was attributed to the World Bank's failure to adapt to the 'small economy' of Kiribati:

The World Bank is traditionally involved in bigger countries. And to come to Kiribati was a very big challenge, because they are doing business in a very small economy, and they were trying to replicate what experience they have from other bigger countries to here … and one thing, in terms of accountability, you know, they are very strict with the finance. But we are saying, you know, you came to Kiribati and this is how we do it. Kiribati is quite free of corruption and all that, but we don't have much capacity. So, maybe if they come to Kiribati with the mindset that we are corrupt, that makes the whole process very different … they don't really have any experience in this kind of place. So, there was bound to be problems, and that is what actually happened.

(Interview with KAP PMU staff member,
South Tarawa, 4 September 2013)

Staffing pressures also placed a burden on project implementation. New positions were created within the Office of the President to support KAP II by providing oversight to the KAP PMU. Yet the positions were initially left unfilled because of staffing pressures in the higher echelons of the public service (Hughes, 2011). Those pressures were exacerbated in the Ministry of

Public Works and Utilities by World Bank policies, preventing consultants involved in project preparation from being involved in implementation. Thus, in the initial years of the project, no extra staff capacity was provided to the line ministries to execute project activities.[6] This situation placed extra pressure on already over-stretched staff, as described by an employee from the KAP PMU:

> There is recognition that the Ministry of Public Works is the local implementing agency for the coastal protection infrastructure, and also some of the water components [of KAP]. But the problem that they have at the Ministry of Public Works is that they lack capacity also. All the people who are employed there have their own job description, and to ask them to do extra was always going to be very challenging.
>
> (Interview with KAP PMU staff member,
> South Tarawa, 4 September 2013)

The combination of over-ambitious objectives, a complicated and fragmented project design, and onerous World Bank requirements was too much for the small KAP PMU to handle. At the same time, among staff in implementing ministries a sense of ennui, pessimism, and perceived failure prevailed, with many redirecting attention from KAP project activities to areas where progress could more readily be made. With no permanent in-country presence, the World Bank offered little support – indeed, the first visit of relevant World Bank personnel to Kiribati was nine months after the project launch (Hughes, 2011). In the words of one KAP II consultant: 'the project was destined to fail'. Expenditure targets soon fell behind schedule. More than one year into what was initially envisaged as a three-year project, nine per cent of the project's total resources had been disbursed. The lag in expenditure alarmed World Bank personnel, prompting a joint donor supervision mission to Kiribati that resulted in KAP II being rated as 'unsatisfactory' and then restructured.

The restructure of KAP II involved rolling multiple contracts into single contracts, contracting management and procurement advisers, and commissioning two international firms to manage the freshwater and coastal protection components of the project respectively. The increased budgetary demands of these decisions were offset by cutting activities and outputs in other areas. The activities cut were mostly 'soft' adaptation measures, including policy and regulatory activities on land use planning, the integration of population policy with climate change adaptation, awareness-raising and capacity-building among communities on the outer islands, and a feasibility study on freshwater lens creation using land reclamation (Hughes, 2011). The fact that soft adaptation measures such as behavioural change and regulation are more difficult to measure, less visible, and arguably more difficult to achieve than 'hard' measures (such as seawalls) likely contributed to this decision (see Fankhauser and Burton, 2011). The advisers managed to speed up the project, but their brief time in-country limited transference of skills

(Hughes, 2011; World Bank, 2011). The requirements and expectations of the World Bank remained unchanged. Operational demands – including efficient expenditure according to project timelines, upward accountability, and creation of concrete and tangible outputs – trumped other more negotiable and time-consuming project goals initially articulated in the Project Appraisal Document, participatory processes among them.

As hitherto demonstrated, the project design for KAP II left much to be desired; nevertheless, some staff in the KAP PMU took the unsatisfactory rating given to the project as personal criticism of their abilities. KAP II was perceived by project staff as failing to reach expenditure targets because of local lack of capacity but, at the same time, the over-ambitious design of the project was perceived to be *causing* local lack of capacity, highlighting the circular, relative, and performative nature of the discourse. This tension was apparent in the language used by an employee of the KAP PMU in describing the reasons behind the restructure:

> So, *we were rated as unsatisfactory* because the measurement was taken that you spend half the money by the middle of the program, but that was not happening. I think we had only disbursed 10 per cent of the total budget. The budget was $6.5 million, if I recall correctly, for the whole project [KAP II]. So, it was difficult to disburse the money *because we lacked capacity, and the reasons for that is that there was that element of an overly ambitious design of the program* [emphasis added].
>
> (Interview with KAP PMU staff member,
> South Tarawa, 4 September 2013)

While this participant in our study evidently concluded that the project's complicated design contributed to the unsatisfactory rating, the weight and power of judgements imposed from outside clearly rankled. Instead of the World Bank being compelled to reflect on whether its procedures and requirements were reasonable and feasible within the Kiribati context, their necessity and legitimacy was taken as given: benchmarks by which to judge who, and how others, lacked capacity. Yet, full compliance with World Bank procedures in the time frames required by the World Bank may never be feasible in small island administrations. Or, if full compliance is realised, inevitably it will be a trade-off against other and potentially more beneficial goals.

If the project was clearly not feasible, why was this not recognised during the formulation of the Project Appraisal Document? This question is especially pertinent given the emphasis on conducting national consultations during KAP I. One interpretation is that, despite extensive consultation prior to the formulation of the Project Appraisal Document, when it came to the point of finalising the document, the World Bank 'dictated' the process and the Kiribati Government adopted a passive stance in response. A staff member from the KAP PMU describes this:

There should be common understanding and consultation … inclusive of everyone, you know. But I think there was no such thing. The World Bank came in and dictated what they think should be the project design. There has to be consensus! Perhaps the government was taking it too lightly, just saying: that's fine, that's fine, that's fine until … it was actually implemented and they realised that there [were] more faults, or more considerations for those activities, and that they should have said 'this is not good, this is not going to be implementable'.

(Interview with KAP PMU staff member,
South Tarawa, 4 September 2013)

Perhaps the Cabinet was under the illusion that the World Bank would provide more support during implementation; perhaps the government was reluctant to say anything that might stop the expected flow of assistance. The World Bank, too, might not necessarily be expected to understand the context for, after all, this attempt was their first to implement a national-scale adaptation project in Kiribati or any atoll nation for that matter; yet perhaps World Bank personnel had a responsibility to inform themselves adequately about the nature of Kiribati before signing off on such an ambitious project.

It is worth noting here that in consultation encounters a number of cultural factors repeatedly disadvantage the I-Kiribati. First, the custom of showing respect to elders and guests makes it difficult for I-Kiribati, including government officials, to confront or contradict foreign development professionals (Donner and Webber, 2014). The expectation that younger people will show respect to their elders also influences interactions between I-Kiribati officials within government; younger people often defer to elders in their presence, even if they have more relevant experience, higher qualifications, or elevated official seniority. This practice can be perplexing for foreign development professionals, who usually give precedence to formal professional hierarchies. Second, shyness, modesty, and fear of sticking out from the group are common cultural traits and powerful behavioural influences among I-Kiribati. Few government officials appear willing to openly question the assumptions of foreign development professionals, as observed by the lead author during several donor–government consultations. Compounding these disadvantages is the fact that consultation sessions usually take place in English, the native language of many Western consultants, but a second language for I-Kiribati.

Finally, there are at play certain power relations between donors and recipients that are also well-documented in the development literature (Cooke and Kothari, 2001; Kothari, 2005, 2006). As donors have their own objectives, consultations are a process of seeking compromise: project actors – including development practitioners and local elite – collude to generate 'planning knowledge' heavily conditioned by dominant interests and by perceptions of what donor agencies are willing and able to deliver, a situation that emphasises short-term tangible outputs and conceals divergent perspectives (Mosse, 2001, p. 23). Although local actors are not powerless in these encounters, ultimately they are expected to reframe development needs in terms that make them fundable. Since donors are not expected to adapt in such ways, negotiations are power-laden and do not represent discussions between equals (Mosse, 2001). A government official from the Office of the President highlighted this imbalance:

> What we see now is that donors come in with a set of priorities and a scope for the program they are doing. So, that's the problem I think. And it has been difficult to try to redirect those funds to areas in country that we see as important as well.
>
> (Interview with government official, Office of the President, South Tarawa, 25 September 2013)

Given that foreign consultants arrive in Kiribati with pre-determined ideas about the scope and types of activities they are willing to fund, it is noteworthy that the official cited above repeatedly sought to steer donors towards activities that intersect with national priorities.

KAP II: unintended and maladaptive consequences?

Perhaps predictably, KAP II had few *outcomes*. Aside from the physical investments outlined in Table 4.1, it did however produce several *outputs* to meet World Bank requirements, among them training workshops, and over 80 reports, mostly written by external consultants (Hughes, 2011). Those reports included a baseline survey of community attitudes about climate change (Hogan, 2008) and a coastal calculator in the form of a spreadsheet intended to assist decision-making regarding coastal infrastructure given inevitable and unavoidable uncertainties and trade-offs (see Donner and Webber, 2014). A similar 'rainfall calculator' was created to assist adaptation decision-making in relation to the installation of rainwater tanks. Additionally, several 'orphan' outputs were produced: preparatory activities that were cut following the restructure, among them a series of outer island profiling reports into which climate change and sea level rise considerations were incorporated. After the geographical scope of the project was narrowed following that restructure, those reports were shelved and the insights from them were not used to inform the implementation of adaptation activities on outer islands (World Bank, 2011).

KAP II was completed in 2011: by 2013, several of the physical investments described above were disused, dysfunctional, or causing maladaptive impacts. The example of the Taborio-Ambo seawall is illustrative of this outcome (Figure 4.2). The seawall joined another and older seawall already causing erosion. Instead of halting the erosion, the KAP II extension of the seawall caused material to shift along the coastline, exposing the water pipeline servicing the population of South Tarawa. Remediating this problem is a priority for KAP III.

Table 4.1 Physical investments from KAP II (2006–11) (US$5.8 million).

Coastal protection works	
Type	*Location*
4 x seawalls	Bairiki-Nanokai causeway
	Korobu (outside SDA HQ)
	Taborio-Ambo causeway
	Airport runway
Mangroves (over 37,000 seedlings)	Makin, Butaritari, Maiana, Aranuka, North and South Tarawa
Freshwater management	
4 x rainwater tank systems at community and public buildings on South Tarawa	Dai-Nippon school, Betio
	Catholic Maneaba, Betio
	Protestant Maneaba, Bairiki
	Government housing, Bairiki
Replacement of tank stand	Nawerewere hospital, Bikenibeu
Water Infiltration Gallery	North Tarawa
Replacement of pumps, pipes and valves	Buota water reserve

[a] A *Maneaba*, referred to in this table, is a traditional meeting house that holds special significance in I-Kiribati culture.

Source: Annika Dean, based on adaptation site visits during fieldwork.[a]

Figure 4.2 Erosion on the coastline adjacent to the Taborio-Ambo seawall, Kiribati. Photograph by Annika Dean.

Rainwater harvesting tanks installed under KAP II also failed to achieve effective and sustainable adaptation outcomes. By 2013, three of the four systems were dysfunctional because of lack of maintenance or vandalism, and broken taps were not replaced; otherwise, tanks were left unused, apparently because of better access to tanks provided by other, later, projects. Damage to tank structures also reduced holding capacity and created habitat for mosquito larvae.

National consultations had been conducted to inform the broad project focus, but outreach and consultation with the intended recipients of physical investments, including the rainwater harvesting systems, was poor. The rainwater harvesting systems were installed on community and public building sites on the assumption that providing tanks to community organisations would be seen as an egalitarian gesture towards climate change adaptation. Yet conflict and confusion typified usage rights and it became unclear who was responsible for maintaining the tanks. In some cases, with no one using the tanks at all, the overall effect was an inverse tragedy of the commons (Hardin, 1968).

More concerning than the superficially maladaptive impacts described above was the suggestion, made by a government official from the Ministry of Public Works and Utilities, that such impacts may run deeper, and that the conditions and requirements imposed by the World Bank and other donor agencies may have amounted to enormous social adjustment, and even loss of culture:

> When the donor agencies come, they say: 'we'll give you the funding for this or that', but then they say 'and you have to do it this way, because that's the requirement of the funding'. Sometimes in the end, the projects don't even meet the original need that we asked for. It's like running in circles, because the donors say 'We've already given you money! Why hasn't your situation improved?' The donors expect that when they give us money, we will do it their way, but we lose part of our culture in that process. Maybe we have to change the way that we operate here. Maybe we have to conform to all those international criteria, but in the process, we lose part of our culture. Something has to give there.
>
> (Interview with government official, MPWU,
> South Tarawa, 19 September 2013)

Discourses that portray islanders as lacking capacity imply that islanders should adapt themselves in sometimes far-reaching and deep-seated ways, detrimentally impacting upon self-image and self-esteem. Additionally, as suggested by the government official above, the conditions and requirements attached to projects, intended to control how projects are executed, can result in real or perceived loss of culture. Kiribati culture is seen to be a core source of strength and resilience among I-Kiribati, enabling people to cope with future challenges including climate change, as exemplified in the quote below:

[The strength of Kiribati is] the people. I feel strongly for what we are doing in this country. The people care. The donors come and go, but we care about what is happening to the country and take it to heart, because we want a future for our children. And our cultural identity is still very much intact compared to many countries, which I think is also a source of strength for our people.

(Interview with government official, MPWU,
South Tarawa, 19 September 2013)

The suggestion that controlling conditions and requirements imposed by donors on recipients can precipitate loss of culture speaks to the need to rethink how suitable or otherwise such conditions are, and to consider may be the wider socio-cultural repercussions of such conditions beyond the blinkered sphere of the project.

At the end of KAP II, the World Bank (2011) revised to a satisfactory rating the unsatisfactory assessment given in the mid-term review. The Bank's Implementation Completion Report recognised that mistakes had been made by both parties and that lessons had had to be learned, but attributed the project's poor performance predominantly to local lack of capacity. The World Bank's mistakes were deflected onto the Kiribati Government, and were framed in terms of the Bank's failure to recognise, and account for, the Government's lack of capacity. There was apparently no critical reflection on its own modes of operation, which were essentially unquestioned. On the Bank's website and beyond (Webber, 2015), KAP II was documented as a resounding success:

The main adaptation investments supported by KAP II not only provided immediate results in terms of reduced vulnerability, but also helped to demonstrate and promote climate risk awareness in planning and design ... KAP III will build on its predecessor's successes to improve climate resilience by both strengthening the government's and communities' capacities to manage climate change effects.

(World Bank, 2013, n.p.)

The combined cost of the preparation and the pilot implementation phases of the Kiribati Adaptation Project was nearly US$10 million. For this sum, few sustainable outcomes were achieved. Instead, significant time and money was spent on World Bank bureaucracy and satisfying World Bank requirements, leaving the Government of Kiribati with nothing to maintain physical investments. The reluctance of donors to provide funding for project maintenance, and the unrealistic expectation that governments will assume responsibility for the costs of expanding 'pilot' projects, is a major reason for the failure to sustain climate change adaptation in the Pacific Islands: 'projects are usually terminated after external funding ends, and the situations commonly revert to those existing prior to the inception of the projects' (Nunn, 2013, p. 151). A comparable situation is found in the 'seawall mindset', the naïve and misguided belief of many donors that hard shoreline structures along Pacific Island coasts are the most effective and enduring long-term solutions to shoreline erosion (Nunn, 2009, 2012). The tragedy of the donor-funded construction of seawalls at mostly iconic sites in the Pacific Islands has spawned an uncritical belief in their efficacy, leading to hundreds of rural communities constructing similar structures, only to see them collapse after 18 to 24 months and eventually be abandoned. The coasts of the Pacific Islands are littered with the remains of collapsed and ineffective seawalls, the legacy of misplaced donor interventions; replanting mangrove fringes would have been more effective and sustainable.

Had the KAP been directed through existing governance systems and via the recurrent budget, the Kiribati Government arguably could have achieved many of the project outputs at a fraction of the cost – and may have raised capacity at the same time. Nevertheless, lack of critical reflection by the World Bank meant that KAP III repeated many previous mistakes. For instance, during the formulation of the Project Appraisal Document for KAP III, the Bank's influence was reportedly overbearing, one government official from the Office of the President describing it thus:

> We were involved in the design phase of the KAP III; we thought it was going to be different from KAP II. But for us, it has ended up just the same as KAP II. You know, the things that we put in as priorities were taken out, and they very much confined the project to the things they wanted to do. So, yeah, it's a bit frustrating.
>
> (Interview with government official, Office of the President, South Tarawa, 25 September 2013)

The World Bank's procurement procedures again were perceived to be inflexible, lengthy, and complex. World Bank requirements stipulated that all local positions in the Project Management Unit should be re-advertised in the transition between KAP II and KAP III. In theory, this step may seem reasonable and harmless, but in practice it was not. Indeed, it took 750 hours or roughly 100 working days over a six-month period to complete the process, as described by the external policy mentor from the Office of the President, whose job it eventually became. This process was reportedly frustrating on a number of levels. The Office of the President had requested the World Bank to allow staff continuity, given the lack of suitably qualified competition; this request was denied by the Bank. According to the external policy mentor:

> The World Bank just doesn't understand that it's not really a competitive market. People are not competing for these jobs that have the qualifications to do them. We are competing with other projects, to recruit the people who are qualified to do the job! That's where the competition is! And they just do not get that. You tell a procurement person that and

they are like, well, no, you have to advertise and if you get three applications, tick that box. But it's always a risk getting new people in from outside to here.

(Interview with external policy mentor, Office of
the President, South Tarawa, 4 September 2013)

The small populations of island microstates, as described hitherto, pose absolute constraints on the number of people with specialised skills qualified to do certain technical jobs, limiting competition for jobs such as those described above. The lack of anonymity in island microstates means that staff of the Office of the President would likely know, or know of, people in Kiribati with relevant qualifications for open jobs. Given that staff in the KAP PMU had been employed for five years, they were seen as qualified and experienced people to continue, but the Bank did not recognise these basic realities of island life. Even more frustrating for the participant quoted immediately above was the fact that none of the time involved in completing the procurement process 'involved formulating climate change policy or formal representation of the Government of Kiribati ... it was all basically admin stuff and meeting World Bank requirements'. Eventually, KAP PMU staff were rehired, with a small amount of natural turnover – two leaving to work elsewhere. Because of the protracted process, staff were also left with a salary gap between the end of KAP II and the start of KAP III, which had to be covered by the Office of the President.

Adopting a nissological approach: a pathway forward?

This chapter speaks to the importance of trying to understand islands and islanders from within – from a nissological perspective – and of moderating development and climate change adaptation approaches to suit a given context. While it is accepted that there are institutional constraints inhibiting donor agencies from completely overhauling their procedures, it is not in the interests of anyone involved if the means thwart the end; that is, if the requirements of development agencies become so cumbersome that they prevent desired outcomes from being achieved and sustained. There seems to be an endemic lack of recognition from donors that some of their requirements and procedures may be impractical in island contexts. After all, what use is efficient expenditure, or even upward accountability, if aid does not address the goals, aspirations, and values of its intended beneficiaries, especially if the consequences of such failures contribute to loss of culture, disempowerment, and low self-esteem?

The Kiribati Adaptation Project clearly suggests a need to devise workable compromises that are functional in island contexts. Devising such compromises requires leadership on the part of the Government of Kiribati in determining clear plans for what it hopes to achieve, and only accepting donor funds that directly contribute to those goals. Accepting funding for funding's sake is likely to result in detrimental outcomes, and restrict the remaining available capacity to pursue priority development and adaptation goals. Devising such compromises also requires the suspension of continental assumptions on the part of donor agencies so that improved approaches can be developed that address their essential requirements, but that are flexible and commensurate with the spatial and socio-cultural characteristics of islands. To build capacity in others requires the ability to build capacity in oneself, which demands humility, self-critique, flexibility, adaptation, and commitment to ongoing learning.

We suggest that adopting a nissological approach in development and climate change adaptation will encourage systems that work for island contexts instead of uncritically imposing norms derived from continental experience. Nissology is less a coherent theory than a philosophical stance. As such, one of our aims has been to unsettle continental assumptions in relation to geographies of development and to demonstrate the power of such assumptions in diminishing island worlds. In this case, we have shown that the continental expectations embedded in donor procedures and requirements in the case of the Kiribati Adaptation Project hindered the project's own outcomes from being achieved and sustained, and may have precipitated more insidious and maladaptive impacts on self-image, self-esteem, and cultural identity. We acknowledge that, for those who have been educated in continental settings, abandoning deeply held continental assumptions is not an easy process, and requires ongoing and diligent critical self-reflection and reflexivity.

Another of our aims has been to prise open discourses of capacity and their instrumental effects. Too often, donors portray islanders as lacking capacity with little elaboration of what is implied by this phrase, and limited explanation of what is expected of islands and islanders. Donors fail to contemplate whether in fact it is they who perhaps need to develop sufficient capacity to be able to understand island-country needs and work effectively with island stakeholders to meet those needs. We suggest that framing capacity in the former (traditional) way functions to stabilise continental discourses and the development status quo, clearing new spaces for donor intervention, enabling donors to blame local lack of capacity when projects fail, and vindicating the *modus operandi* of donors. This pattern lends legitimacy to the adoption of controlling approaches by donors, despite rhetoric about consultation, participation, and recipient-country ownership. We posit that islanders themselves are better positioned to determine notions of capacity in their geographical and socio-cultural contexts and with regard to their own aspirations and values, and that donors should look to their own methods of engagement for explanations as to why so many of their interventions have failed to be sustained.

Notes

1 The term islandness is preferred to insularity as it emphasises the relationship of islands with the sea, 'island' being derived from the old English word 'ealand'

meaning 'waterland' (Neemia-Mackenzie, 1995, p. 1). By contrast, insular is defined as: 'the state or condition of being an island', connoting isolation and remoteness (Oxford English Dictionary, 2015, n.p.).

2 Other models include PROFIT and SITE. PROFIT economies are characterised by their ingenuity, flexibility, and active domestic policy used to promote economic diversification. PROFIT economies exploit their amorphous jurisdictional status to develop tax and insurance havens, offshore banking centres, and duty-free exports. SITE describes Small Island Tourist Economies that have used the idyll of remoteness to promote tourism as an engine of their economies (Oberst and McElroy, 2007).

3 Most donors to the Pacific have benefitted from Pacific resources, currently or historically: Australia has benefitted from the phosphate reserves of Nauru and Kiribati; Japanese aid is given where fishing access is provided and withdrawn where it is restricted, while Taiwan and China use aid to compete for diplomatic recognition in the Pacific (see Atkinson, 2010; Teaiwa *et al.*, 2002; Wesley-Smith, 2013).

4 The RERF is the sovereign wealth fund of the Government of Kiribati, established by the British Colonial Government in 1956 with royalties from phosphate mining. The RERF functions as a budget support mechanism to buffer against revenue volatility, and a form of asset backing for government borrowing (Purfield, 2005).

5 'Ecosystem Based Adaptation uses biodiversity and ecosystem services in an overall adaptation strategy. It includes the sustainable management, conservation and restoration of ecosystems to provide services that help people adapt to the adverse effects of climate change' (Secretariat of the Convention on Biological Diversity, 2009, p. 6).

6 Providing labour to implement the project was part of the Government's in-kind contribution, as agreed by the World Bank and the Government of Kiribati. Nevertheless, the reluctance of donors to fund labour to implement projects is a significant hindrance to implementation and drain on existing capacity.

5   An island feminism

Convivial economics and the women's cooperatives of Lesvos

Marina Karides

Islands give expression to particular sets of gendered and sexual arrangements that deserve consideration by feminist geographers and sociologists, island scholars, and other social scientists: this is because of, for example, their apparent isolation; ecological particularities; connections among visitors, tourists, experts, or kin; long-standing transnational social networks; distinct forms of militarism; or locally enduring communities. This chapter presents an empirical application of island feminism, a perspective I develop to bridge feminist and queer insights and island studies thought. My intentions are to enrich the critical perspectives brought to bear in the study of island places, and to examine social inequalities on islands. Nuanced understandings of islandness depend on appreciating and assessing the gendered experiences and social and spatial organisation of island communities. An island feminist perspective can also expand the scope of feminism, benefitting future strategies for island policies – including those that address economic development or climate change – by recognising and supporting the varied gendered strategies needed to maintain island livelihoods and preserve island topologies.

Island feminism refers to the intellectual sensibilities of island place and constructs of gender and sexuality, positing them as intertwining forces that shape the particular conditions of economic, social, and ecological life, and the cultural and political machinations particular to islands. Island feminism also reappropriates narrow understandings of 'island women', highlighting the agency, geographical awareness, resourcefulness, and forms of community shaped by categories of islandness and marginalised by constructs of sexuality and gender. Like most feminisms, it is action-oriented in pursuit of just and fair conditions for all beings, and is guided by specific interests in local and subaltern strategies that remain resistant to hegemonic discourses and practices of power. In island feminist research, a range of issues is examined in terms of culture, place and space, and identity, among them enterprises, collectives, and environmentally attuned practices created and lived by marginalised populations in marginalised places, such as certain groups of women on certain islands. I conceive of these enterprises, forms, collectives, and practices as *convivial economics* (Karides, 2012).

With few exceptions, island studies has not included gender and sexuality in considering the meaning of the term islandness (but see Barney, 2007; Lattas, 2014). *Island Studies Journal* and *Shima*, for example, presently have little in the way of theoretical and empirical research considering how island societies and culture shape, and are shaped by, the social construction of gender and sexualities. At the same time, feminist or queer research on islands seems to give little acknowledgement to island specificities. In response to such gaps, and drawing on field work and interviews conducted on Lesvos, Greece, between 2008 and 2012, I offer a case study of 11 of the women's cooperatives that exist on the island, and do so in order to read into existence a nissological feminist geography. A long-standing tradition on Greek islands, in general terms cooperatives provide a means for earning an income collectively – in ways based on full ownership of an enterprise by employees, democratically governed by members, and serving and supporting communities (Hacker, 1989; Jessop, 2001; Miller, 2006). In each of the women's cooperatives, those I met with and interviewed primarily produce and sell local and traditional foods. They are located in villages of different sizes throughout Lesvos, an island with low mountain chains and startling seashores that is a summer destination for returning kin, Northern Europeans, and tourists and travellers who prefer small and local experiences over those offered in large-scale resorts and hotels.

Lesvos is part of the Aegean island chain that separates mainland Greece from mainland Turkey and it sits just eight kilometres (five miles) off the Turkish coast (39.1° N, 26.3° E). Like many archipelagos, the Aegean islands are liminal locales that serve as 'borderlands' between social constructs of place and space such as East and West and, at the same time, that are cultural amalgams of regions they are seen to separate. These islands hold the social and cultural histories of Ancient Greece and the Byzantine and Ottoman Empires, which had lengthy periods of control over these islands. In many instances, food items produced by the women's cooperatives reflect regional rather than national cuisines, and are influenced by indigenous or native foods and local agriculture, as well as by those imperial histories.

Both the enterprises crafted by the women's cooperatives and the reasons that explain their genesis speak to island economic strategies that secure financial well-being and provide for the survival of island communities. The study of everyday life on islands offers scope to assess mundane work and varied economic forms, practices, and transformations that otherwise can be eclipsed by social sciences' prevailing emphasis on institutional structure, quantifiable patterns, macro-economic processes, or governance and policy (Bourdieu, 1998; de Certeau, 1984; Lefebvre, 1984). In turn, an island feminist approach draws on feminist understandings of gender, sexuality, and work, and highlights how everyday life on islands provides grounds for distinct economic practices centred in place. In this sense, it is noteworthy that both islanders who remain *in situ* and retain cultural and economic traditions, and others who leave mainland locations to create livelihoods on islands, often choose to engage in small and local economies that are community-centred or ecologically based rather than participate in profit-seeking expansionary goals and material accumulation. Thus, enterprises on islands may be construed as subaltern, and those that are created by groups marginalised by gender, race, or class, or who choose to adopt subaltern practices, seem to demonstrate a double resistance to neoliberal development (George, 2010; Robinson, 2014).

On such understanding, in the following sections I review the literature on feminisms, place, and economic development; the subaltern qualities of island place that support economic alterity; and the growth of alternative economics. I introduce the idea of convivial economics, and then present a thematic analysis of data I collected on the women's cooperatives in Greece. The timing of my research coincided with Greece's economic crisis, which has had severe consequences for the nation, especially in Athens, and which has centred on employment and the cost of living. Most of the women's cooperatives have been able to survive the economic challenges arising from the austerity measures imposed by the International Monetary Fund, the European Central Bank, and the European Commission – measures that may have aggravated the crisis in Greece rather than assuaged it (Borooah, 2014; Monstiriotis, 2011). The chapter concludes by considering the resiliency of women's cooperatives and convivial economics, and by reflecting on their relevance to an island feminism framework.

Defining convivial economics: island practices in gendered space

Feminisms, place, and economic development

Feminisms are essential frames of analysis through which scholars and others have identified the narrowing of economic opportunities, limits to power, and claims to space according to gender. Although feminisms have generally centred on women because they have been most marginalised by gender categorisation, recent work has identified 'intersectional' patterns and practices that perpetuate the privilege of some groups while limiting others. It was Black feminist thought that reoriented feminism by highlighting the intersectionality of gender with other social categories (Collins, 1991). In other words, opportunities and experiences are gendered and simultaneously are shaped by race, class, age, or nation (Choo and Freree, 2010). Island feminism builds on the intersectional approach, introducing islandness and island sense of place as significant social and spatial qualities that influence work, communities, social networks, and families.

Contesting mainstream geographic thought in the late 1970s and 1980s and critiquing the literature for 'excluding half of the human in human geography' (Monk and Hansen, 1982, p. 11), feminist geographers have drawn their own subfield, bringing to the fore gendered analyses of space and place (Hanson and Pratt, 1995; Johnson 2008). Over the course of several decades, feminist geographers have since promulgated the view that 'space is gendered and gendering has profound consequences for women' (Doan, 2010, p. 1; McDowell, 1999; Rose, 1993). Engaging in various methods, feminist geographers have described the processes that gender and sexualise space, for example in relation to urbanisation (Massey, 1991b), economic development (Safa, 1995), or tourism (Besio *et al.*, 2008). Feminist geographers have also emphasised micro-level analyses to articulate how gender relations occur in daily contexts affecting politics and ideas about household, families, and the body (Massey, 1994). Feminists' initial focus on women was later elaborated to conceive of gender and space more broadly. As Gibson-Graham (2006, p. xxiv) explains, a 'feminist spatiality embraces not only a politics of ubiquity (its global manifestation), but a politics of place (its localisation in places created, strengthened, defended, augmented, and transformed by women)'. Thus, economic development, particularly in the Global South – and including many islands – has been one of the key foci of feminist research. Both feminist sociologists and geographers have roundly critiqued traditional strategies of economic development and Western proposals of modernisation. The patriarchal norms embedded in modernisation strategies are known to have overlooked women as workers for industrial development. And women's work in postcolonial societies, often in small local enterprises, has been dismissed as traditional and either inconsequential or anathema to development.

Researchers working in the field of gender and development have noted that women often use small-scale and informal means to make ends meet when formal wages are insufficient or unavailable, thereby subsidising capital accumulation (Freeman, 2000; Harrison, 1991; Hsiung, 1996; Mies, 1986). Often shut out of formal sector jobs due to discrimination, but needing flexibility and income (including second and third sources of it), women have relied on their creative energies and resourcefulness to build small-scale enterprises, production systems, and services. Researchers have also highlighted how the gendered division of labour, or the household and caretaking work largely assumed by women, has supported the development of enterprises that require them to work double duties (Isserles, 2003).

Although feminist scholarship has highlighted women's agency in the construction of micro-enterprises, many times these efforts have been erroneously reduced to forms of coping or viewed as reactions to dire economic circumstances and gendered constraints (Mullings, 1999). Yet self-employment or collective means of income-earning by women extend beyond coping and often are generative – they do not simply yield to the constraints of larger forces. Drawing on these alternative, traditional, and sometimes innovative sources of income and community development, women reject formal, hierarchical, and punitive labour. Instead, they create work that can sustain autonomy, offer better labour conditions that highlight interpersonal engagement, value familial responsibilities, and improve their communities (Karides, 2005, 2007, 2010).

In many instances, smaller economic operations have developed organically and are based on traditional small-scale local economies that have resisted, or have been overlooked by, large-scale economic developments. Island societies characteristically have maintained small enterprises embedded in particular social and ecological contexts. These types of enterprises, particularly cooperatives, have often gone unseen or been ignored by those eager to modernise using large-scale and industrial development strategies, as well as by researchers (Geertz, 1963; McClelland, 1967). Yet, as Hau'ofa (1993, p. 2) has recognised, 'academic and consultancy experts tend to overlook or misinterpret grassroots activities because these do not fit in with prevailing views about the nature of society and its development'. More than two decades on, the persistence and renewal of cooperatives and other forms of alternative economy suggest that locally grounded economic activities are resilient and offer methods of economic exchange that continue to counter the neoliberal ethos. These small-scale and local development strategies have attracted consideration by island studies scholars. In turn, island feminism draws attention to the small, locally owned, and environmentally conscious enterprises that are characteristic of islands. These types of enterprises, cooperatives and collectives, farmer markets, small and micro-businesses, and informal distribution services, have been vehicles for marginalised groups to create community, and have also been a means of survival. Enterprises such as cafés, bars, small stores, and other operations created by and for marginalised groups provide the means to contribute to local economies, and they permit informal social and political discussions and exchange to occur among those without access to privileged spaces – here, then, is a novel provisioning: a means to create spaces of one's own.

The subalterity of island place

Unless 'anywhere is possible', the geographic dimension of alternative economics requires further consideration (Leyshon and Lee, 2003). Where and how cooperatives and alternative markets sprout up also reflects what spatial parameters or challenges exist in relation to them in any given location. Island studies perspectives, such as those advanced by Hay (2006) or Baldacchino (2006, 2008), assume that islands hold distinct sets of economic, social, and ecological conditions, and those perspectives inform my own argument that there is need to consider the spatial or geographical possibilities of economic alterity on islands. Indeed, recent work in human geography helps to mark the existence of subaltern spaces, defined as marginal locations in which develop certain orientations oppositional to mainstream or dominant economic and social practices. Definitions of spatial subalterity are often mapped out by radical geographers who seek to understand places in which people and communities do not subscribe to global logics of capitalist accumulation (Jazeel, 2014). Clayton (2011, p. 246) distinguishes subaltern space in a double sense as:

... a space of denial, denigration, and exclusion, in which the subaltern inhabits and is assigned a place in society, and granted an identity, that marks her/him as subordinate, a discrepant place, or diverging and cut off from, yet constituted by, elite authority and its norms; but also as an alternative and counter-hegemonic space in which the desire and ability to fracture and challenge power is imagined and enacted.

Because the production of space is geographically and historically contingent, it is in local sites that counter-discourses can be established – those that have not been grasped by apparatuses of power or that refuse to acknowledge particular kinds of power. Yet, in a growing body of alternative economic literature considering the expansion of local, collective, or sharing economic entities, more than one perspective neglects to examine the spatial imperative for alternative economic enterprises. In addition, others discount the notion of local or alternative economics built around attachments to place, conceiving them as parochial or separatist (Gibson-Graham, 2006; Hess, 2009; Wright, 2010). These views often seem to stem from privileged outposts or are based on macro-level frames of analysis. Harvey explains:

The depiction of others' geographical loyalties as banal or irrational ... helps foster ignorance and disinterest in the lives of those others; meanwhile space after space is opportunistically demonized or sanctified by some dominant power as a justification for political action. Such biased geographical knowledges, deliberately maintained, provide a license to pursue narrow interests in the name of universal good and reason.

(Cited in Clayton, 2011, p. 246)

Clayton's (2011) central thesis is that subaltern space is paradoxical and includes phenomena that produce dominant modes of economic practice and daily existence, *and* ways of being and doing that have never adhered to mainstream logics and practices or that are created as alternative ontologies in the midst of these contradictory forces. Some islands may be understood as subaltern spaces. Such islands often remain at the fringe of mainstream developments in which space is conceived as 'an anywhere' – as interchangeable, absent of particularisms, and without diversity. Indeed, because of their particular and diverse geographies and histories, islands are fertile grounds for convivial economics to flourish. In this light, a gendered spatial subalterity helps to explain the women's cooperatives of Lesvos and their practices of convivial economics as strategies of place, informed by location and by historical and contemporary patterns and rhythms of engagement that inform social systems and challenge inequalities in them.

Convivial economics

For generations, small-scale, local, and community based enterprises – here termed convivial economics – have been created by island women and others groups whose lives are grounded in subaltern spaces. Research on alternative economics, such as that by Laville *et al.* (2007) or Leyshon and Lee (2003), has provided only some insights on where these subaltern enterprises exist and persist, including in relation to localism (Hess, 2009), social economy (Wright, 2010), and community economics (Gibson-Graham, 2006). In addition, the fact that gender and sexuality shape economic choices, including alternative economic choices, is all but absent in much of the alternative economic literature (Allard and Matthei, 2008).

By placing the margins and marginalised at the centre of things, an island feminism focused on convivial economics draws together an intersectional and placed-based understanding of alternative economic practices. In practice, convivial economics is based on negotiation rather than regulation; facilitates autonomy, community, and sociality; and may overcome or undo the constraints of a dominant economic order. Convivial economic practices often exist alongside large-scale developments – for example, side by side with corporate tourism operations on islands.

In developing an island feminist framework of economic practice on islands, I draw from Illich's (1973) work, *Tools of Conviviality*, and Schumacher's (1973) *Small is Beautiful*. Illich's critique anticipated the current growth of alternative economic practices and studies, and is directed against both capitalist production and forms of socialism in which workers are not in direct control of, for example, hand tools, factories, and systems of decision-making. According to Illich (1973, p. 35), the alternative, encapsulated by the idea of conviviality, is framed by three values – survival, justice, and self-defined work: economies with 'tools that guarantee the right to work with independent efficiency'. In turn, Schumacher (1973) argued that small-scale operations, no matter their numbers, are less likely to be harmful to the natural environment than large-scale ones. Simply due to scale, he argued, the environment could more easily recover from the impact of multiple small encroachments than from single large ones. Like Illich, Schumacher emphasised the use of correct technology and affordable, accessible, and fitting approaches to small-scale application and creativity.

Illich's and Schumacher's works invite further consideration of the future of island development in ways that support economic activities oriented to sustaining place, environment, and cultural forms and practices as people negotiate social change. Their arguments are additionally relevant here because they reflect many of the economic practices found on less-developed or remote islands, not least among enclaves ecologically, economically, or socially compromised by large-scale tourist developments – Lesvos being a case in point. In the following section, I provide some background on the development of cooperatives in Greece, and then turn to a thematic analysis of the findings from my engagement with the women's cooperatives.

Cooperatives in Greece

In the nineteenth century, the cooperative movement across Europe was broadly conceived in response to the inequities and harm caused by advancing capitalist industrialisation (Wright, 2010). In particular, agricultural and consumption cooperatives developed as responses to such inequities and to harms felt by small producers, and were intended to maintain their livelihoods as large-scale and industrial farmers infringed on their livelihoods (Moulaert and Ailenei, 2005). In other words, those who formed cooperatives were resisting and challenging the growing *scale* and intensification of production along with its exploitive conditions, and were constituting subaltern spaces in the process.

The International Labour Organization (ILO, 2007, n.p.) defines cooperatives as 'an autonomous association of persons united voluntarily to meet their common economic, social and cultural needs and aspirations through a jointly owned and democratically controlled enterprise'. In general terms, the organisational forms that typify cooperatives have varied on the basis of the size of an enterprise and the product or services provided; however, the cooperative form generally implies full ownership by employees and democratic government and governance by members (Wright, 2010). Hacker (1989) presents several ways to measure the extent and depth of cooperatives' commitment to these precepts, including the degree to which workers participate in decision-making, have access to information, and share profits.

In the development of an alternative economy, cooperatives are now being revisited as a new form of governance that depends neither on the logic of accumulation nor state control (Jessop, 2001). For the many who work in and advocate for them, cooperatives are seen to provide a stable, humane, and egalitarian method of production. Beyond the commitment to ensure fair and safe labour, cooperatives depend on solidarity and a mutually constructed vision of the enterprise being engaged in. They challenge hierarchical models of management, and members seek to support benign social, cultural, and environmental systems (Miller, 2006).

Across Europe, cooperatives are now expanding again such that one in four new enterprises is thought to be part of the social economy; for example, social cooperatives increased five-fold in Italy between 1993 and 2000. There and in France, Belgium, Spain, and Portugal, state laws have been passed better defining cooperatives, particularly social cooperatives, and recognising them as an expanding movement (European Social Commission, 2011). In Greece, as in other parts of Europe, cooperative forms of production date back to the nineteenth century: fishers, shepherds, artisans, and farmers worked collectively, distributed income equally or by reference to the nature of one's work, and collaborated on financial decisions, sometimes with the aid of elected boards (Petropoulo, 1993). Some of the original cooperatives, such as those in olive oil production, have continued for more than a century.

Since the middle of the twentieth century, Greek cooperatives, and particularly larger agricultural ventures, have sometimes been subject to heavy-handed state control. Agricultural cooperatives have often been used to create political patronage and secure food supply for growing urban populations. During Greece's period of dictatorship from 1967 to 1974, cooperatives were forced to abide by the demands of the regime, producing directly for the state. In the 1980s, under the influence of the first elected socialist party, PASOK, agricultural cooperatives were expanded and government invested in them.[1] In the 1990s, Greek cooperatives multiplied in diverse sectors, independently orienting themselves nationally and to the global economy (Nasioulas, 2012). In more recent times, as extreme austerity measures have been imposed on Greece, and strikingly high unemployment continues, cooperatives have increased in proportion to gross domestic product, and have been formed mostly as grassroots efforts to curtail unemployment and provide affordable access to basic goods. Only in the last few years has the Greek government taken formal account of the social economy through legislation and policy generally referred to in terms of social economy and social enterprise (Nasioulas, 2012).[2] Legislative reform in Greece is part of a growing commitment to social economic development promoted by the European Union and the growing independent network of cooperatives in the region.

The women's cooperatives of Lesvos

The women's cooperatives of Lesvos are found in villages across the island, and may be read as a heterogeneous and organic form of resistance to neoliberalism. For instance, in all the cooperatives, members pay themselves the same rate regardless of position; wages are based on hours worked. Like other Greek cooperatives, these collectives demonstrate endogenous economic cultural practices in which community and hospitality are linked to a decisive preference for economic independence. Table 5.1 lists the names of the villages, the years the cooperatives started, and the number of members in each. Cooperatives are found in small and quiet villages tucked along mountain ridges and valleys, and in larger settlements such as Molyvos – a highly touristic and architecturally fantastic village, or Agiassos – located in a mountain landscape rich with wild herbs, olive groves, and chestnut trees that bustles with religious tourism. Most cooperatives are found in the *agora* or the local marketplace, which is identified by street signs even in the smallest of villages.

Coming from their nearby homes and arriving one by one, often on foot, to production and retail facilities, in the morning, or later in the afternoon following a traditional and extended break, women's cooperatives members across Lesvos consider their schedules and reorganise them in response to regular last-minute requests for goods. Preparing pastries, savoury treats, or fruit preserves, women constantly engage with each other, rely on one another to make production plans, and trust each other to decide when to

Table 5.1 Lesvos women's cooperatives: village, year initiated, and membership.

Village	Year initiated	Membership
Agra	1998	15
Anemotia	1998	10
Asomotos	1998	4
Agia Paraskevi	1999	11
Ayiassos	1998	12
Filia	2004	11
Mesotopos	1998	34
Molyvos	2004	13
Parakoila	2001	16
Polichnitos	1997	9
Petra[a]	1982	Unknown
Skalochori	2000	17

[a] Although I met with the Petra women's cooperative, highly regarded for its excellent restaurant, it operated more as a family business than a traditional cooperative.

Source: Karides, 2012.

work overtime, discussing current events and politics over cups of Greek coffee. They ply and fold thin layers of dough, sometimes spread over ten feet on a table between them, and manipulate the dough into traditional artistic patterns and shapes flavoured and embellished with local nuts, cheeses, and honey. In summer, almost all the villages in which women's cooperatives operate receive kin returning from Athens for vacations, as well as Greeks and foreign tourists visiting the island; many of the latter will stop and visit cooperatives to sample foods. Then, from sometime in September to sometime in April, villagers live in almost complete isolation.

In 2008, 2009, and 2012, I held audio-recorded interviews in Greek with members of each of the 11 women's cooperatives on Lesvos, and took extensive field notes from ethnographic observations during visits to both their production and retail facilities. I also considered ephemera such as cooperatives' pamphlets. In analysing findings, I used grounded theory as a constant comparative approach, documenting emerging categories while coding data and then comparing codes in and between categories, continuously developing and sorting categories, integrating them, or abandoning them in order to produce thematic, conceptual, or theoretical knowledge from social research (Charmaz, 2006; Glazer and Strauss, 1967; Strauss and Corbin, 1990). My conversations with cooperative members grounded me in the realities of daily life and the concerns held by island women in Lesvos. In what follows, I present insights from those conversations, organised as five themes that serve to flesh out the island feminist perspective I seek to advance.

Out of the house and into the agora

Gendered spatiality shapes the sociality and enterprises found in the *agora* and the *platia*, or public squares, of Greek island villages. Certainly, past studies on gender in Lesvos highlight the significance of spatiality in the construct of gender and sexuality in Greece (see, for example, Loizos, 1994). Conducting research on Lesvos in the 1980s, Papataxiarchis (1991, p. 157) noted that the:

> most remarkable aspect of social life is the extensive segregation of the sexes: women and men spend most of their time in same-sex contexts and identify strongly with their own sex. The nucleated village seems to be divided into sex-specific territories. The square in the centre with the coffee shops and marketplace is dominated by men. The houses that surround the square and constitute neighborhoods are small havens for women and children.

In large part, and with the exception of Mytilini, the island's capital, village life in Lesvos maintains the gendered spatial code that Papataxiarchis (1991) described. In fact, unless they are contributing to a family enterprise, shopping, or collecting children from school, the women of Lesvos are generally not seen socialising in public or in village *kafenions* (coffee shops). Indeed, public space is viewed as less virtuous than the home and it is seen as appropriately inhabited by men. Traditionally, women found in the *platia* are stigmatised for appearing to be sexually available (Zinovieff, 1991).

  In the course of my research I routinely drove to and through many villages on the island, making my way to the various women's cooperatives. Relative to men's presence, the completeness by which women were absent from sitting together at coffee shops on the *platia* was striking. Of the more than 20 villages I passed through and visited at various times of day, invariably only groups of men were visibly collected together. Only during summer were those gendered spatial practices somewhat loosened.

  However, women's cooperatives have challenged the norms of gendered public space in Lesvos's villages by claiming space in the *platia*, and for many with whom I spoke, engaging in collective work that brought them outside the home was among the reasons for joining a cooperative. Gianoula, a member of the Agia Paraskevi cooperative, bluntly stated, 'the house isn't enough; it chokes us'. The cooperatives were a vehicle for village women to develop both economic enterprise and modes to engage socially in the public sphere. In most cases, cooperatives are housed in buildings in the *agora* in and around the *platia*. Almost all of the cooperatives combine production facilities with a shop-front and a small outdoor or indoor café. For the first time, women have captured a portion of the village centre where they can convene and drink coffee, albeit in conjunction with their work efforts. As a senior member of the Molyvos cooperative explained, 'we can't stay at home – we have learned to work. I can't stay at home and watch the TV all day. We take great interest in our work, and the development of the cooperatives; it exhausts us, but we love it.'

Arguably, then, on Lesvos the isolation of the domestic sphere and of island village life are being challenged by a focus on convivial economics, and are giving rise to a nascent island feminism. In one example, when exploring how a cooperative got started in Agra, I was told by one woman: 'rain, a fireplace, and coffee'. The Agra cooperative is located on the Kalloni–Eresos road, the only thoroughfare on the southwestern side of the island, and one that hugs the coast as it makes its way down from the mountain town. The population of Agra is just under a thousand. The cooperative members explained that if they were going to continue living in their village they needed a project. They shared with me how the public middle school closed, and voiced concerns over closures of the local elementary school as villagers departed for Athens. After many rainy winter afternoons and sessions of brainstorming, the Agra cooperative started, the women using their own funds and kitchen equipment.

In another example, in the fall of 2008 at a public meeting of the Prefecture of Lesvos and Limnos in Mytilene, I was witness to the discussion of a matter related to the women's cooperatives. A government-sponsored business known as the Lesvos Shop was threatening to close – one that promoted the island's goods and in which many of the women's cooperatives placed their products for sale (and by 2012 it had indeed closed, indebted to many of the cooperatives). It was fairly late, around 10:30 pm, when members of the cooperatives had the opportunity to voice their concerns to the Council, gathered in a large meeting hall. Meri, a cooperative member of Ayiassos explained to the President of the Prefect, Council members, and a few hundred members of the public: 'the economics is one thing, but it is getting out of the house, getting to be free. I threw away my mental health drug. It is not just economic to work with each other it is for our health, to maintain tradition and the community ... the cooperatives are everywhere.'

To stay in their island place and to stay healthy in it, women's cooperative members thus have recognised the need to challenge traditional spatial gender norms and to develop projects that extend them beyond household managers. Members of various cooperatives indicated that their collective work did not replace their domestic duties – instead they managed both, and in just a few cases women members felt that their communities shunned their efforts. Although I could not verify it, they suggested that men with enterprises in the *agora* identified them as competition. Yet, what mattered for the women was getting out of the house, participating in the marketplace, and working with others. As one member explained, 'we have learned the routine of work. What are we supposed to do now, return to our houses?' Along with the benefits of being publicly engaged, women spoke positively of the relationships they built with each other, and they valued their collective enterprise. In this sense, although the members of women's collectives claimed space in retail locations in which they served regular customers in the *agora*, they operated in homo-social environments that seem to characterise a significant part of Greek life. I witnessed their intimate ease, the synchronicity with which they conducted their work, and the physical closeness in which they operated. As Georgia, a member of the collective in Agia Paraskevi explained, 'we have a big love for each other, when one of us is missing, we look for her'. Another cooperative member explained how their children grew up playing together as an extended family.

In addition to camaraderie, friendship, and public engagement, most of the cooperative members highlighted the value of engaging in creative enterprises and artistic projects. As a member of the Asomotos women's cooperative explained, 'we have artistry, we want to see beautiful things. We have dreams for our future and we love what we do. This isn't something that we are made to do. We love our life.' Artistry is apparent in cooperatives' products, packaging, and displays, as well as in the exactness by which deeply local recipes are prepared. Skilfulness is highly valued, too: producing orzo pasta with repetitive hand and finger movement, or shaping perfect roses from carefully mixed almond paste, demonstrate the adeptness of cooperative members. Rather than focusing on individual skill, then, the cooperatives also highlight the pleasure of collective production and many of the women credit their economic success to that practice.

Survival and success in the cooperative model

The survival and success of the women's ventures was attributed to the cooperative model. The Mesotopos cooperative offers a case in point – one showing how convivial economics embedded in place can succeed. The Mesotopos cooperative comprises 34 members, 15 working on a daily basis and others casually.

In 2012, the Mesotopos cooperative moved into a large new building. Using their own funds for capital, the members bought land directly on the main road, and constructed a shop. The first floor houses a spacious room tiled throughout and with lovely display cases. In those are decoratively placed traditional Greek pastries such as *kouroubia, blatzites, deeples, flouxeria, baklava, koulourakia*, and cookies in various shapes basted with butter, sometimes rolled in sesame seeds or drizzled with chocolate. Set atop the counters are carefully made pyramids of jars filled with colourful preserves made from quince, olives, watermelon, cherries, oranges, lemons, apricots, and more. There is a large stacked display of *gliko tou koutaliou* or 'dessert of the spoon', a dazzlingly sweet dessert served in small spoonfuls with coffee. Entering the shop, one is met with shelves of multi-coloured homemade pastas in various shapes, each labelled with production and expiration dates written by hand.

The Mesotopos cooperative has been recognised for its achievement with an award by the national women's agriculture association. When I asked members about their success, one explained:

We are a cooperative; we all make decisions together and we have achieved our success because we have maintained the cooperative model. We all come together and make our decisions and agreements. We decided together to purchase the land and agreed on the plans for the building, we took out a loan. We began the efforts for the new building three to four years ago.

The emphasis on collective decision-making and support was characteristic of all the cooperatives. For example, a member of the Anemotia cooperative observed that a 'cooperative is unique. We can work within ourselves. To owe, to borrow, we can fight, and you can start your battle. We see the unemployment out there. We make our plans, begin our project, and start our battle.' A woman from the Molyvos cooperative explained the same dynamics in these terms: 'the cooperatives can survive because collectively we can agree to reduce our wages, a small businessperson who hires someone is still responsible for paying their wages'. In a strikingly similar statement, a member of the Polichnitos cooperative offered the following: 'the cooperative has more strengths, we can agree to make cuts to our salaries, and work longer hours; when you have someone working for you have to pay them. Here we decide together to make cuts. We fight besides each other and together to maintain it.' In short, the cooperative model allowed the women's cooperative members to engage in self-defined work (Illich, 1973) and democratically decide how to adjust incomes when necessary or to develop production. They valued the use of an economic and political organisational form that had enough flexibility to consider the various needs of the members. Although in a few cooperatives the same woman was president in 2008 and 2012, offices were usually decided by biennial elections.

Women's work: selling culture and place

For many of the women's cooperatives a key source of income comes from villagers and kin now living in places such as Athens and Thessaloniki, who purchase prepared foods and other items made in their own specific village cooperatives with their own unique recipes. In Athens, the Peri Lesvou shop in the Monastiraki neighbourhood is dedicated to the sale of products from Lesvos. The shop carries foods from all the cooperatives (including weekly deliveries of fresh yoghurt in customary clay pots), and when I met with the owners in 2012 they confirmed that many of their customers are connected to Lesvos. Equally important sources of income for the cooperatives are the weddings and baptisms on Lesvos and further abroad (notably as far as Central Africa and Australia) where migrants from Lesvos reside. When I asked some of the women's cooperative members about their export businesses, Ourania from Skalochori explained that a 'private bakery could never provide the quality of product that the women cooperatives provide and the villagers know that'.

There also seems to be a far-reaching attachment to *women's* products, a niche that the cooperatives clearly fulfill. Traditionally, village women prepare local foods for major life events and those women now living in cities may not engage in these rituals of food production for such events, such duties consigned to the distant women's cooperatives on Lesvos. Yet, city women help sustain island identity by purchasing the products exported to cities by the women's cooperatives and that indirectly helps to maintain unique cultural forms that keep their villages alive. Paradoxically, the women cooperatives both preserve and challenge traditional and gendered spatial norms, while protecting the specificities of island place.

Caring for topos and community

The women's cooperatives of Lesvos seem steadfast in their commitment to place and community. They identify with the countryside surrounding their villages and rely on local agricultural products and wild-harvest fruits and herbs to create their products. While there are many ingredients typically associated with Greek foods such as honey, nuts, olive, citrus, and feta, the cooperatives all draw from specific contexts and places to produce goods unique to their locales. For example, the village of Ayiassos holds a seasonal chestnut festival, and the women's cooperative is central to it. What seems distinct about island place is attachment to particular parts of an island or a specific island village in which identities are invested. One member in the Ayiassos cooperative, located in a region noted for cherries and chestnuts, elucidated:

> When we started it was not just about making money. For us it was to preserve our land and for ourselves. The economics of the project was not our priority. We wanted to create a project that was outside our work in the household or in the fields. It is our commitment to our team, to each other, and to the importance of this project for our village, and we think Lesvos at large that we have been able to preserve despite the setbacks handed to us. If the pursuit of a project is primarily or only focused on money then it cannot succeed.

The village of Parakoila offers another example of these dynamics of identity, place, and convivial economics. Most of the village's families arrived during a population exchange in 1923 between Greece and Turkey, after the Catastrophe in Smyrna and the end of Greek efforts to expand through Asia Minor (known as the Megali Idea). Cooperative members there explained to me how their grandparents moved into homes vacated by 'Turks' who had resided in the village, and clearly the architecture in this particular village is influenced by Eastern or Ottoman design, a remnant of an old Mosque visible from the main road. Yet, while 'Greek', cooperative members nevertheless identify with Eastern roots, the President telling me: 'we are Anatolian, our kitchen – the pastries derive from Eastern recipes. This is what we love to do. Our cuisine comes from Micro-Asia.' Here, the confluence of Micro-Asian cuisine with Micro-Asian architecture has a distinct effect on visitors, and possibly on residents, and the cooperative's fare 'fits'. In this sense, Parakoila presents an interesting case to examine islands as borderlands and hybrid and subaltern spaces.

Convivial economics and local gendered development

In the summer of 2012, I asked members of the women's collectives to tell me about the effects of the global economic crisis. One in Parakoila, who was usually less talkative than others, captured a widespread approach to the crisis by reciting the adage 'when you are soaking wet, it doesn't matter how much more it rains, you are already wet'. Other members of the cooperative nodded, and this attitude seemed typical. When I raised for discussion the status of the Greek economy, referring to loan agreements and austerity measures, or to the Memorandum of May 2010 between Greece and 'the troika', members rebuffed my inquiries with a look or simply stated something that has the same meaning as 'the show must go on'. Cooperative members were committed to working hard, taking pay cuts, or withholding pay in order to secure the survival of their businesses. Many had used such strategies during other periods of financial distress, and were resolute that the current crisis was another bridge to be crossed. During my visits, they were constantly engaged in production, with continuous orders to fill. There was no indication that business slowed and, as in 2008, my interviews and meetings often had to be rescheduled as I sat and watched women jump to fill a last-minute order. Indeed, many cooperatives were engaged in plans to grow their enterprise and attract new customer bases for their goods. Cooperatives with younger members, such as that at Parakoila, considered themselves more innovative and as one member stated: 'we are constantly thinking of ways to expand our market. Cooperatives with members in their fifties and sixties make items as they have. They are not thinking about a new customer base, or understand the new interest in traditional goods.'

Predictably, an arena for future development was tourism. As Tessie, another member of the Parakoila cooperative, put it: 'there is room for expanding the tourism we offer. In Greece we get the tourist and throw him in the pool, we take him for a walk to the beach so he can also have a swim in the ocean, and then bring him to a taverna at night.' Tessie was drawing attention to the cultural attractions and the quiet getaway feeling that remote islands can offer urban tourists. Others in the Parakoila kitchen added, 'we have so much to offer here on the island of Lesvos'. The President of the cooperative went on to discuss with me a recent visit the cooperative had hosted involving personnel from a Greek cooking show, 'Gourmet Alive', that they believe contributed to interest in their products. They were conceiving of ways to attract upscale culinary tourists to their village. As a member explained, 'we want to make our place known'.

Yet, the President of the Molyvos cooperative, which is highly reliant on and experienced with *foreign* tourists, was less optimistic about that industry, noting that 'when you are waiting for tourism, others make the plans; you don't make the plans. For now we are just trying to keep the place open. We are cutting our wages. We are here all day long.' In 2012, Molyvos seemed especially affected by Greece's economic crisis: the protests in Athens and the media's portrayal of them had discouraged cashed-up foreign tourists. Most other cooperatives, more dependent on purchases by community members and kin living in large urban centres, were then better placed. Certainly in some cooperatives, an alternative strategy was to focus on wealthy Athenians who might seek a remote location and learn about local cuisine while vacationing. Several cooperatives sought to capitalise on growing interest in local economies or foods. A few suggested that the survival and even success of the women's cooperatives throughout the worst of the economic crisis was due to a growing pride in localism. Lina, a member in Agia Paraskevi, explained this trend in the following terms:

> I think now many people are reading labels and trying to figure out if they came from Greece. This is a new consciousness. We [Greeks] have been importing without developing export. But Greece has a great deal that it can grow and produce and we have to organise and exploit it, [and then] there would be less need and less debt.

These discussions with members of the women's cooperatives informed and support an island feminist perspective. The women were conscious of their island places and what it took to stay and to make livings and lives in those places, and they knew how to secure personal well-being, and support their communities and their land. While challenging long-held norms about space and place that circumscribe gender and sexuality in remote island villages, they have also upheld deeply rooted local traditions that adhere to food production and collective enterprise. In the terms offered by one woman, Sasa,

> an island is a wonderful place to come and to relax, but living here it can also strangle you. If you live on a mainland, the thought to leave your circumstance, or to escape, is there even if it may not be a real possibility. On an island you are a captive of the ocean. So you have to birth wings to get past it. For us the cooperative is those wings.

Conclusion

This study is the first that theorises gender and islands together using the terms of island feminism, and it both draws attention to how economic forms on islands contrast the hegemonic discourses of development and considers how they are informed by gender, sexuality, nation, and ethnicity; it adds islands as one of the phenomena that shape social worlds and spatial dynamics.

Island feminism considers the numerous rhythms that come together in island places and that contribute to economic practices preserved or formed by groups whose members have not been well served by the dominant economic system. Island economies are more likely to hold informal and negotiable ties and relations and locally driven small-scale enterprises because they are contained spaces, and yet economic exchanges that emanate from islands and that are embedded in island social relations also stretch out across space and thus challenge tropes of islands as isolated. The existence of the women's cooperatives of Lesvos and their economic and social engagements with village, island, and mainland populations, places, and enterprises surely support the idea that islands and island regions are globally or regionally interconnected, for example by long historical ties, diasporas, and exchanges with mainlands and across archipelagos.

The convivial economics I found in Lesvos may also capture the Mediterranean's 'own ways of opposing capitalism' (Leontidou, 1990, p. 2). Organised in local contexts, organic economic alterity flows from modes of production and exchange that have not been defined or coopted in hegemonic discourses and practices. Through their collective economic endeavours, the women's cooperatives create convivial work environments that are supportive and sociable, that build island communities, and that demonstrate small-scale, local models of development that reject dominant logics of economies that are profit-driven and based on endless expansion. The cooperatives show how women otherwise or hitherto marginalised by certain norms of gender and sexuality create meaningful work that challenges those norms while protecting communities' well-being and upholding cultural resilience in island places.

Postscript

Greece and the island of Lesvos have recently taken centre stage in the media landscape, the former because economic crisis, the latter because of the immigration crisis. For the residents of Lesvos, immigrants arriving on rafts that have crossed the often choppy strait from Turkey are not new. I witnessed the lack of regard for life jackets on the shores of Eftalou in 2008. Villagers from that part of the island have used these and the rafts to make insulation for their homes. The convivial economics practised by the women cooperatives have extended to numerous local efforts, grassroots and unseen by the mainstream media, to aid these new arrivals. The ethos of care and work that is embedded in one's own environment or one's 'shima' (Suwa, 2007) with collective goals, or convivial economics, might be a path to consider both for surviving a volatile global economy and for supporting refugees and immigrants in new livelihoods.

Notes

1 At the same time, the State supported the development of women's agricultural cooperatives as part of PASOK's interests in increasing tourism in the Greek islands. Presently, there are 140 such cooperatives in Greece, and a yearly conference is held in different regions, during which there is recognition of the best cooperative for the year (Lassithotaki and Roubakou, 2014).

2 In passing, members of the women's cooperatives spoke to me about some of these new policies and legislation during my visit in 2012. Concerned that the changes would prove a hindrance, for example by requiring a minimum number of members in cooperatives, most saw the laws as a hurdle they had to address to conduct their business: none saw those changes as advantageous and a few decided to ignore them.

6 Nature and islands

Rethinking the cultural heritage of New Zealand's protected islands

David Bade

Introduction

In whatever form cultural heritage may take – a structure, a building, an archaeological site, or a memorial – it is a key societal concern, revealing who we have been, where we have come from, and who we are. Although the word 'heritage' and its attendant ideas have existed for centuries, only from the 1980s did it become the focus of intensive academic inquiry (Smith, 2012). Heritage is now widely acknowledged to mean anything from the past that is considered important enough to be handed on and conserved as a legacy for future generations (and that includes the positive, the undesirable, the grand, or the ordinary).

Alongside heritage scholars and environmental historians, human geographers were among the first to engage with the study of heritage. David Lowenthal's (1985) *The Past is a Foreign Country* was a landmark work for heritage studies. Since then, human geographers have studied heritage in terms of key themes such as identity and collective memory (Graham *et al.*, 2000; Johnson, 2004); power (Alderman, 2003; Hardy, 1988; Tunbridge and Ashworth, 1996); sense of place (Titchen, 1996; Waterton, 2005); economics and tourism (Johnson, 2009); and locality and scale (Harvey, 2014). Human geographers have also questioned forms of binary thinking that, in protected *natural* areas, can lead to detrimental results for *cultural* heritage items, sites, or precincts, which may be disregarded, dismissed as insignificant, or even be removed because they are not deemed 'natural'.

Several such studies have focused on islands (Bade, 2010, 2013; Carter, 2010; Cronon, 2003; Head, 2000; Hennessy and McCleary, 2011; Olwig, 1980; Wockner, 1997). In New Zealand, the focus of this chapter, island reserves are lifeboats for indigenous wildlife (Butler *et al.*, 2014). Throughout the country, islands have become refuges for threatened native species, sites of ecological restoration, and places where exotic mammalian and plant pests have been exterminated using pest control and eradication techniques. New Zealand's island reserves are considered 'natural' or at least are seen as 'becoming natural'. Because of this mind-set, the cultural heritage of these islands is often deemed of secondary importance. For example, little attention is paid to Māori histories of land use, and their settlement on and association with the islands; and European settler farming histories are repeatedly considered insignificant, even though they reveal the use of the islands for logging, quarantine stations, and military purposes. Given the foregoing, this chapter is an inquiry into the development of the idea of 'islands as natural' in New Zealand, which has prevailed since European settlement. By reference to colonialism, national identity, the nature/culture dualism, heritage, and the geography of islands, the chapter presents a discussion about how islands have become central to New Zealand's nature conservation efforts, and considers what impacts such mentality has had on the management of cultural heritage on New Zealand's island reserves.

Natural heritage and New Zealand's islands

In 1891, the world's first island nature sanctuary was established on Resolution Island in Fiordland in the southwest region of New Zealand. The sanctuary was primarily for the protection of two endangered native flightless birds: the kakapo (*Strigops habroptilus*) and the kiwi (*Apteryx*). During the same decade, three other island sanctuaries were established: Secretary Island, also in Fiordland, in 1893; Little Barrier, in the northern Hauraki Gulf, in 1895; and Kapiti Island, west of the lower North Island, in 1897 (Figure 6.1).[1] In relation to both their geographical materiality and imaginary, islands play a significant role in natural heritage conservation in New Zealand.

Islands' geographical qualities, conditions, and characteristics – such as separateness, boundedness, isolation, vulnerability, and smallness – heighten their 'naturalness', which has influenced European attitudes and approaches to the New Zealand landscape and to the Māori since the mid-nineteenth century. That colonial period brought to New Zealand new people, plants, animals and, just as importantly, new ideologies. As geographers have shown, the tendency to separate nature and culture, borne out of the Enlightenment in eighteenth-century Europe, was significant and powerful. The division of 'nature' and 'culture' into two realms of reality was neither straightforward nor universally accepted (Anderson, 1995; Dingler, 2005; Olwig, 2009). Yet this framework spawned interlinking environmental transformations and attitudes to the environment that led to the establishment of the aforementioned island sanctuaries, and positioned New Zealand's islands as central to nature conservation activities from which culture appeared absent. Over a hundred years later, New Zealand is still leading the way in terms of island ecological restorations involving the removal of exotic pests and weeds, the establishment of native forest, and the (re)introduction of threatened native species. At the 225-hectare (555-acre) Karori Wildlife Sanctuary in Wellington, New Zealanders also pioneered, in 1995, the concept of 'mainland islands' – lands surrounded by a natural or artificial border which are ecologically restored (Star, 2014). Since then, numerous other mainland islands have been created.

Figure 6.1 New Zealand showing the location of most of the islands and water bodies referred to in this chapter. Source: Bade, 2013, p. 63.

The environmental transformation of New Zealand was initiated by Polynesian settlement from the late thirteenth century (Wilmshurst *et al.*, 2008). The early burning of the land by the Māori and the associated arrival of the Pacific rat (*Rattus exulans*) resulted in the elimination of 50 per cent of both pre-settlement forest areas and the late Holocene assortment of bird species in New Zealand, including the moa (a giant flightless bird, *Dinornithidae*) (Anderson, 2002; Haggerty and Campbell, 2009).

With the arrival of Pākehā (non-Māori and predominantly European peoples), New Zealand's environments underwent further transformation; native forests were cleared by means of logging or burning, and over two million hectares (five million acres) of native bush was cleared between 1886 and 1909 (Star and Lockhead, 2004). Some 85 per cent of New Zealand's original wetlands were drained by World War II. Fire was used to clear much of the tussock lands of the South Island for pasture (Haggerty and Campbell, 2009). Indeed, most of New Zealand's landscape was converted from forests and wetlands into pasture land for farming between the mid-nineteenth and mid-twentieth centuries (Brooking *et al.*, 2002; Brooking and Pawson, 2011). Exotic flora and fauna from Britain and mainland Europe – many of which are now considered pests – were both purposefully and inadvertently introduced (Isern, 2002). Agriculture was also undertaken on New Zealand's islands, ranging from subtropical and volcanically active Raoul Island in the north to sub-Antarctic Campbell Island in the south (Bellingham *et al.*, 2010).

In New Zealand, as in other British colonies, in the middle decades of the nineteenth century first-generation European settlers were more concerned to transform landscapes into something that resembled their homelands rather than to preserve what originally existed (Gentry, 2006; Ginn, 2008; Young, 2004). For most, the loss of indigenous forests was a necessary cost of European settlement and yet, as the burning and clearance of New Zealand's forests continued, the impulse to preserve also emerged. Advocates of the national parks movement promulgated the idea that it was important to reserve places of nature as distinct, distant, and protected from culture and civilisation. Borne primarily out of the United States in the 1870s, the movement was enthusiastically taken up in New Zealand. The first national park there, located at Tongariro on the central plateau of the North Island, was established in 1887, only 15 years after the establishment of the world's first such park in Yellowstone (Baird, 2013; McLean, 2000; Thom, 1987).

The formation of national parks and scenic reserves in New Zealand was closely associated with feelings of nostalgia for, and, indeed, a sense of regret about, the destruction of, New Zealand's native forests. This emotional geography was particularly apparent from the late nineteenth century as Māori and second-generation Pākehā remembered a time before areas of land were cleared for farming. Certainly, by the late nineteenth century natural scientists were documenting the varied impacts of environmental transformation on New Zealand's biota (Treadwell, 2005).

Scientists voiced their concerns – about the diminishing numbers of native flora and fauna, and particularly the loss of bird species, as a result of fire and land clearance; about the collection of bird skins for overseas institutions; and in relation to the presence of mustelids such as stoats (*Mustela erminea*), weasels (*Mustela*), and ferrets (*Mustela putorius furo*), which had been introduced to control rabbit populations. Scientists around the world, as well as from New Zealand, became fascinated with New Zealand's unique and declining forests and animal species, such as the kiwi, kakapo, takahe (*Porphyrio hochstetteri*), and tuatara (*Sphenodon*). Such research and interest helped the national government and public to appreciate that New Zealand had something of international significance worth protecting (Star and Lochhead, 2002). These reorientations in attitude mark the emergence of an idea that New Zealand had natural heritage both unique and ancient. This revelation was entwined with the concept of ecological climax underpinning scientific thinking at the time. Accordingly, every biotic community was expected to reach a state of climax that was stable and in balance unless disturbed. The notion of climax perpetuated the idea that humans are separate from nature and have detrimental impacts upon it (Cronon, 1995; Griffiths, 1996). To scientists, humans could be a 'contaminating' disturbance to nature's balance and stability. The idea that 'reserves' could protect nature from the impact of human activity came to be advocated among scientists as a means to prevent the loss of further native biota.

Isolated and separate from threats present on the mainland, New Zealand's offshore islands came to be seen as ideal sites to protect species. However, in establishing island reserves the national government and scientists failed to register – or simply ignored – the occupation of and authority over those islands by Māori, for whom islands were not a separate category but an extension of land reached via what had been their principal means of transport, the waka (canoe). All offshore islands in New Zealand were in some way used or settled by Māori (Davidson, 1990). Despite this Māori history and its palpable cultural heritage, islands were seen by members of Pākehā society, and by those in the science community in particular, as ideal natural refuges for endemic species threatened, or not found, on the mainland (Bellingham *et al.*, 2010; McSaveney, 2009).

The island nature refuge in New Zealand was first envisaged at an Australasian Association for the Advancement of Science (AAAS) conference in early 1891. George M. Thomson, a naturalist, and A.P.W. Thomas, Professor of Natural Science at Auckland University College, moved that 'in the interests of science, it is most desirable that some steps should be taken to establish one or more reserves, where the native flora and fauna of New Zealand may be preserved from destruction' (in Galbreath, 2002, p. 84). Thomson and Thomas requested that the national government reserve both Resolution Island and Little Barrier Island. The AAAS secretary, Professor A. Liversidge of the University of Sydney, forwarded the request on 16 March

1891 and the New Zealand government reserved one, Resolution Island, on 22 May 1891 (Hill and Hill, 1987).

Resolution Island was considered highly suitable as a bird sanctuary, providing habitat for flightless birds, such as kakapo and kiwi, and protecting them from mustelids, which were thought to be defeated by the island's water-bounded geography. There were faint hopes that the island may have some takahe, which were nearly extinct at the time (Hill and Hill, 1987). Richard Henry was appointed curator of the island in 1894, and he made some of the earliest translocations of birds. With the help of a well-trained (and muzzled) dog, and possessing expert knowledge of bird habits, Henry managed to trap over 700 kakapo and kiwi and transport them to the island (Peat, 2007). However, in 1900 tourists on a boat close to the island reported that they had seen there a stoat chasing another flightless native bird, the weka (*Gallirallus australis*) (Young, 2004). This observation was later confirmed (McSaveney, 2009). Henry was understandably distraught that mustelids had managed to swim to the island and had started to prey upon the highly vulnerable native birds he had translocated, and the event illustrates a major problem in the management of exotic species invasions that continues to hamper in-situ conservation on island reserves.

Other islands were also suggested as possible bird sanctuaries. In August 1891, journalist James Richardson presented to a meeting of the Otago Institute a paper entitled 'On the Extinction of Native Birds on the West Coast'. The ensuing discussion generated suggestions for potential island bird sanctuaries in addition to Resolution Island (Hill and Hill, 1987). Among those suggestions were the aforementioned Little Barrier Island as well as larger islands or eyots in Lakes Te Anau, Manapouri, and Wakatipu; Stewart Island; Codfish and Bench Islands (close to Stewart Island); the Snares (south of Stewart Island); the Solanders in Foveaux Strait; the sub-Antarctic Auckland Islands; and the subtropical Kermadec Islands. Those present at the Otago Institute meeting stressed that any island reserve must be at a distance from a mainland sufficient to prevent mustelids swimming there. Such emphatic suggestions highlight the importance attributed to islands for nature conservation in New Zealand at the time, their relative isolation and boundedness by water chief among the characteristics of islandness most valued in this regard.

Three more island sanctuaries were established between 1893 and 1897. Secretary Island, New Zealand's fifth largest offshore island, was designated an island reserve in 1893 (Nightingale and Dingwall, 2003). The island remained free from any introduced grazing or browsing mammals until the 1960s, when wild deer became established (Edge, 2004). Little Barrier Island had long been considered suitable as a nature sanctuary. In fact, the island was considered more appropriate than Resolution Island as it had a warmer climate and was less accessible to mammalian predators that could swim (Young, 2004). The island had also attracted much scientific interest in respect of the flourishing native bird populations. However, the Māori presence on Little Barrier Island complicated the establishment of a reserve. Ngati Wai chiefs, such as Rahui

Te Kiri (1830/1?–1930) and her spouse Te Heru Tenetahi (1826/7?–1923), were opposed to leaving the island so that a reserve could be established, and Tenetahi was profiting from milling the island's kauri (*Agathis australis*) and from running pigs and cattle on the island (Young, 2004). Nevertheless, in order to protect the potential reserve the national government issued an injunction to prevent the felling of timber on the island in December 1892 (Hill and Hill, 1987). Two years later, the Little Barrier Island Purchase Act 1894 was enacted, so that the island could be compulsorily acquired from them for £3,000. The *New Zealand Parliamentary Debates* 1894 show how the Bill was robustly discussed.

Tenetahi believed the Act to be in breach of the Treaty of Waitangi (the founding document of New Zealand, signed by the British Crown and leaders of many Māori tribes). Despite objections, at dawn on 20 January 1896 Tenetahi, Rahui Te Kiri, and other inhabitants on the island were forcibly moved to Devonport in Auckland by a bailiff, policemen, and soldiers (Ballara, 2010). The Auckland Institute then managed the island until 1905 when its administration was transferred to the Tourism Department (Young, 2004). Te Kiri and Tenetahi continued to demand compensation for the removal of their people from Little Barrier. Tenetahi refused to remove his stock from the island until 1897, when the Crown decided to enforce that work to recoup some of its expenses. While Tenetahi removed the animals, he also dismantled the only habitable cottage on the island to make it difficult for personnel from the Auckland Institute to stay their during periods of inspection (Ballara, 2010). Regardless, the forcible removal of Māori inhabitants from their island homeland meant a people's cultural heritage and customary rights had been 'sacrificed' in the face of nature conservation.

The fourth island reserve established during the decade was on Kapiti Island in 1897. The island had been the location of New Zealand's largest single area of lowland coastal forest free from introduced predators and herbivores, until possums (*Phalangeriformes*) were introduced in 1893 (Young, 2004). Like Little Barrier, Kapiti Island had many Māori inhabitants, the Ngati Toa, who were unwilling to sell their land to the national government, which delayed the establishment of the reserve. The island had been an important base for Te Rauparaha (c.1760s–27 November 1849), chief of the Ngati Toa, to wage battles against other tribes in the Musket Wars in central New Zealand in the early decades of the nineteenth century. The island was also a key site for interaction and trading between Māori and whalers, with up to 2,000 people living on the island during whaling times (Hill and Hill, 1987). Once whaling declined in the mid-nineteenth century, farming became an important industry and much of the island's forest was cleared. In 1897, the national government passed the Kapiti Island Public Reserve Act, declaring the island a reserve for the flora and fauna of New Zealand. The Act compensated European owners, and allowed about 526 hectares (1,300 acres) of land at Waiorua Bay on the north-eastern side of the island to be used by the Ngati Toa.

Although significant deforestation continued in New Zealand, by the early twentieth century the significance of indigenous natural heritage had been reinforced by the preservation of native bush remnants, including the establishment of island reserves. Thus, to the settlers, native flora and fauna gradually became symbols of national identity. Native birds and mountains appeared on New Zealand's postage stamps as early as 1898 (Star and Lochhead, 2002). The huia (*Heteralocha acutirostris*) and the kiwi became national icons (Ginn, 2008). Likewise, scenic landscape painters found mystery and romance in New Zealand's remaining native forests and represented them as timeless and primeval (Eldrege, 1991). New Zealand's wildlife and landscapes had become designated forms of natural heritage important enough to be passed on to future generations. By the 1930s, islands featured prominently on the list of scenic reserves; in 1934, for example, New Zealand had around 950 such reserves, covering a total of 300,000 hectares (Young, 2004). Those reserves included many islands such as the bulk of Stewart Island; three small islands in the Marlborough Sounds; the Poor Knights Islands; Taranga Island off Whangarei; Raoul Island in the Kermadec Island group; the sub-Antarctic Auckland Islands; and Karewa Island in Tauranga harbour.

From the 1970s and 1980s onwards, New Zealanders then began to acknowledge human impact on the environment and many became determined to preserve what remained – some suggesting that this impulse became embedded in New Zealanders' psyche (Nathan, 2009). In 1972, New Zealand had one of the first 'green' parties in the world, the Values Party (Belich, 2001; Bührs and Bartlett, 1993). The environmental group Greenpeace became popular in the 1980s, especially after 10 July 1985, when the French Secret Service bombed its ship, the *Rainbow Warrior*; berthed in Auckland Harbour, the ship was just about to sail to Moruroa in French Polynesia to protest against French nuclear testing at the atoll. Now, the apparent naturalness of New Zealand has become so fundamental a part of national identity that it is showcased to the world (Kirby, 1996), not least in relation to tourism, natural heritage (Bell, 1996; Peart, 2004), and ongoing scientific research related to ecological restoration, which has proliferated on offshore islands since the 1980s.

Ecological restoration is the process of assisting or accelerating the recovery of an ecosystem which had previously been altered by human interference (Jackson *et al.*, 1995). The notion of 'ecological purity' has become central to such actions (Lowenthal, 1997, p. 238). The ideal of restoration is to 'return' the landscape to some apparently 'pure' and 'original' – that is, pre-human – condition (Eden *et al.*, 1999). However, returning landscapes to such seemingly pristine states means that cultural remnants and relics are reduced to a 'disturbing stain of "progress"' (Lowenthal, 1997, p. 236). Consequently, the preservation and protection of cultural heritage are often considered to be in opposition to ecological restorations.

There are four main reasons for the spread of ecological restoration projects in New Zealand. First, from the 1970s, members of the environment movement fuelled concern for the conservation of natural heritage and improvements to the state of the environment. Ecological restoration was a way to actively engage in the conservation of nature (Eden *et al.*, 1999). Griffiths (1996) writes that advocates of the movement distinguished themselves from antecedents by supporting the removal of indigenous, colonial, and postcolonial layers of history in order to restore earlier 'ideal' times.

Second, and related to this view, in the late twentieth century scientists again began to highlight the significance of offshore islands as repositories of significant biological wealth and as capable of providing a 'direct view of pre-human nature' (Daugherty *et al.*, 1990, p. 18). New Zealand's offshore islands were especially attractive for ecological restoration projects because of their suitability as refugia for many native species, especially those characterised by high levels of endemism (Bellingham *et al.*, 2010). In addition, a growing body of literature on the environmental histories of New Zealand highlighted the detrimental impact humans have had on its forests, wetlands, and bird life (Arnold, 1994; Crosby, 1986; Flannery, 1994; McKinnon, 1997; Park and Potton, 1995). Confirmation of the scope and scale of these impacts has meant there was, and is, public demand to preserve and enhance island ecosystems (Towns and Ballantine, 1993).

Third, concern about New Zealand's natural environment coincided with advances in technology that allowed environmental managers to address detrimental past actions highlighted by historians, scientists, and environmentalists. Such advances finally facilitated the aerial administration of poison baits, for example, which – while controversial in itself – enabled the removal of pest mammals on large islands. Threatened species could also be much more easily translocated to and from islands (Bellingham *et al.*, 2010; Department of Conservation, n.d.). Such technological changes have enabled a shift in approach from preservation and protection of natural heritage to the active management of flora, and particularly fauna, in order to restore and enhance island ecosystems. No other geographical formation in New Zealand has offered the same opportunities for conservation (Towns *et al.*, 1990). Both physically and psychologically, islands are well-suited to ecological restoration programs and designation as natural heritage sites. Indeed, the physical form of an island means that islands can be viewed as places that can be wholly controlled by humans, or equally, totally submitted to nature. In this way, because of their boundedness and isolation, islands seem insulated from developments on mainlands and therefore seem more natural (Bade, 2013; Kelman, 2007). Doubtless and more prosaically, where islands are proximate to larger mainland areas, it seems easier to remove plant and animal pests from them – and thus maintain ecological integrity more readily than might be the case on larger landmasses or, indeed, aquatic environments. After all, islands are signified by their bounded and separate landscapes and by the sense that they are contained and able to contain other matter (Towns *et al.*, 2009).

Fourth, the establishment and natural heritage bias of the Department of Conservation also influenced the growth in number and scope of island ecological restoration projects. For example, although there is a long history of appreciation of New Zealand's indigenous natural heritage, a major shift in approach from the late 1980s has been the idea of converting (restoring) good and valuable pastoral land to its perceived original and unsettled state (Bade, 2010).

The Department of Conservation was formed in 1987 as a central government department to promote the conservation of New Zealand's natural and historic heritage. However, the agencies that formed the department appear to have privileged natural heritage, and positioned the Department to become the chief advocate for nature conservation in New Zealand's protected areas at the expense of a concomitant focus on historic heritage conservation. Managers and staff at the Department made it their role to establish and manage ecological restoration and nature conservation programs on offshore islands from the 1990s (Bellingham *et al.*, 2010). Indeed, as a result of the Department's work, New Zealand is widely considered to be at the cutting-edge of island restoration projects internationally. It was from within the Department that staff actively experimented and tested new eradication methods such as the helicopter bait spread system. From the 1980s to the 2000s, there were numerous translocations to island sanctuaries of threatened native bird species, such as saddleback (*Philesturnus carunculatus*), takahe, and kakapo, as well as the eradication on islands of exotic mammalian pests such as cats (*Felis catus*), goats (*Capra aegagrus hircus*), rats (*Rattus*), mice (*Mus*), and possums (Bellingham *et al.*, 2010).

As a result of these four factors, island ecological restorations have gained widespread public support and enjoyed significant political traction with similar programs throughout New Zealand. This backing and popularity for ecological restoration has largely brought about the formation of numerous community groups – mainly in the form of trusts – in partnership with the Department, and these tend to have the primary goal to ecologically restore landscapes. Currently, the department manages over 70 island ecological restorations in New Zealand (Department of Conservation, n.d.).

Cultural heritage and New Zealand's island reserves

The previous discussion has highlighted how particular colonial ideas about nature gave rise to the establishment of nature reserves in New Zealand; positioned natural heritage at the forefront of New Zealand's national identity; and, notably, placed islands as conceptually and physically central to the safeguarding of New Zealand's natural heritage using strategies to foster ecological restoration. Attention now turns to focus on the implications for the cultural heritage of island reserves of the idea that islands are 'natural' and that aspects of cultural heritage do not belong, or are not significant enough, to be in apparently natural settings. The discussion first concentrates on Rangitoto, an island reserved in 1890 for its natural features, and then on Motutapu, an island undergoing ecological restoration since the mid-1990s. These two case studies in Auckland's Hauraki Gulf exemplify how the idea

Figure 6.2 Rangitoto as seen from Auckland, New Zealand. Photograph by David Bade.

of 'islands as natural' has been made manifest in terms of reserves and sites of ecological restoration.

Rangitoto Island

Around 550 years ago, Rangitoto Island (38.1° S 175.2° E) emerged from the sea in a major volcanic eruption. The substantial volcanic figure of the island now dominates the skyline of Auckland harbour (Figure 6.2). It is Auckland's youngest – and largest – volcano. Despite being only 25 minutes by ferry from downtown Auckland, from the summit the island seems a world away from urban life. From a distance, the welcoming greenness, uniformity, and naturalness of the island disguise its rugged scoria and its human history.

The perception of the island as a place of nature has dominated Rangitoto's history since European settlement. Rangitoto has been considered as distinctive and separate from Auckland; a place that should be protected from the growing city – an antidote to the worst exigencies and effects on natural environments of burgeoning urban life in Auckland, among them land use change, deforestation, and species loss. Rangitoto, in comparison, became a site of significant importance to scientists – and particularly botanists – from New Zealand and around the world, who marvelled at the island's vegetation, growing as it did in what seems like an inhospitable landscape (Wilcox, 2007). Rangitoto also and increasingly became a popular recreation destination for picnickers and day-trippers. Consequently, in mid-1890 the island was scheduled as a Recreation Reserve under the Public Reserves Act 1881

(Cottrell, 1984; Woolnough, 1984). This designation as a reserve is significant in that it was very early – even for New Zealand. The island was valued as a place of nature to be enjoyed by members of the public – and this mirrored developments elsewhere, such as in Australia, Canada, the United States, and the United Kingdom as the parks movement developed and took shape in both the populist imagination and the regulatory frameworks of varied governments.

In the face of mounting pressure from visitors to have overnight stays on the island, for there to be more visitor facilities, and to maintain and develop recreation infrastructure on the island, the Rangitoto Island Domain Board allowed temporary structures to be erected for overnight stays from circa 1910. Despite being illegal under the Public Reserves and Domains Act 1908, this development was encouraged on the basis that it provided a source of income to fund recreational facilities on the increasingly popular island. During the late 1920s and early 1930s these temporary structures evolved into more permanent structures and over 100 baches (basic holiday homes) were built, facilitating the advent of vibrant summertime communities (Yoffe, 1994).

However, between the 1890s and 1930s advocates for greater conservation of the island's natural heritage found voice, their concerns part of the aforementioned growing national regard for natural (native) heritage. Scientists, conservationists, academics, and members of the general public heavily criticised the ways in which developments on Rangitoto – not least the construction of structures such as baches – were destroying the native flora and fauna of the island and undermining its fundamental natural values. One of the most active and forthright advocates for conservation was the Auckland botanist Lucy Cranwell, who delivered a public address in December 1931 on the value of Rangitoto to botanists and of the negative consequences of settlement on the island. Woolnough (1984) writes that, at roughly the same time, Sir Arthur Hill, Director of Kew Gardens, insisted that Rangitoto should remain a nature sanctuary and he condemned the introduction of exotic plants to the island by bach owners. Many letters to the editor encapsulate the desire to maintain the island as a natural place. One, written in pseudonym by 'Settler', was published in the *Auckland Star* in 1935, emphasising the drastic environmental and scenic change on Rangitoto over a 40-year period. In 1895 'Settler' had visited Rangitoto for the first time and later reflected about having experienced 'Nature undisturbed, unadorned by the hand of man' (in *Auckland Star*, 1935, p. 6). Forty years on, in January 1935 'Settler' visited again to find a 'great attempt at modernisation with motor roads, unfinished tennis lawns and fancy buildings' (ibid.). 'Settler' believed that the island 'should have been left in its natural state, otherwise we are destroying a valuable national asset known all over the world' (ibid.). The letter drew on nostalgia about past environments on Rangitoto and sustained the notion that human interference in natural places was negative. In April 1937, and under the influence of such vocal concerns, New Zealand's first Labour Government decided that no new bach leases were to be issued, existing leases were to expire after 20 years, and any additions or alterations were prohibited. This decision meant the bach structures were 'frozen' in their 1937-state (Brassey, 1993; Treadwell, 1994). Yet, in the 1950s, when the leases were due to expire, bach owners successfully lobbied the national government to extend existing leases for the lifetimes of the owners. Over 100 leases were issued at that time and, in a few cases, bach owners placed the lessee title in the name of a minor in the family, thereby allowing a longer lease.

Then, in the 1970s and 1980s, as bach owners died and leases expired, the baches were demolished by the Department of Lands and Survey, then charged with managing the island. Rangitoto was seen as a 'natural place' to be experienced by day-trippers, a state that could not be achieved if it contained any trace of baches or boatsheds (Treadwell, 2005). In 1980, Rangitoto was re-designated from a Recreation Reserve to a Scenic Reserve. This reclassification further emphasised *scenic* natural features and values in opposition to any remnant cultural heritage on the island. From the 1980s, however, there were signs that baches were being perceived not just as holiday homes, but also as heritage. Woolnough (1984) in particular brought to others' attention the historical significance of the holiday bach community on Rangitoto and questioned the rationales then governing the management of the island. Woolnough (1984, p. 68) wrote that by employing the policy of removing the baches once the lease expired, 'we commendably preserve our flora and fauna' yet by doing so we also 'destroy our human history by stamping out all signs of habitation': rightly or wrongly, people have lived on Rangitoto for almost 80 years and we should not 'try to pretend that they never existed' (p. 68).

When the Department of Conservation was established in 1987, it took over the management of the island. By that time, only 33 of the original 140 baches from 1957 remained (Treadwell, 1994). Then, during the 1990s, debates ensued between those who wished to remove the baches and restore the natural appearance of the island and those who wished to retain the baches as reminders of 1930s' vernacular architecture and as tangible expressions of an under-appreciated aspect of New Zealand's social history. As a result of those debates, the remaining baches are now protected, and some have been restored (Bade, 2013). Nature conservation, however, has remained central to the island's management. For example, in 1990 a 1080 poison blitz involving the aerial distribution of poison over land was carried out to rid Rangitoto and neighbouring Motutapu of possums and wallabies; and in 2006, a three-year pest eradication program for the islands was instigated to rid the island of all remaining animal pests.

Motutapu Island

Motutapu Island neighbours Rangitoto to the northeast (35.1° E 174.0° S), and is just a 35-minute ferry ride from the central business district of Auckland. Motutapu's history covers the full span of New Zealand settlement by Māori and Europeans, and has some of the best-preserved evidence of early Māori settlement from the fourteenth century, of European farming from the mid-nineteenth century, and of military activity during World War II. Once part of the super-continent of Gondwana, Motutapu is geologically ancient; has fertile soil excellent for cultivation; and, as a consequence of European settlement, is dominated by pasture grass, with little remnant native vegetation. Motutapu therefore does not have the same sense of rugged naturalness as Rangitoto, with its extensive pohutukawa-dominated forest and distinct volcanic landscape. Yet, and notwithstanding its long human history and extant cultural heritage values, there are strong pressures to engage in ecological restoration on Motutapu. In the early 1990s, restoration plans were made by ecologists and biologists; however, these neither fully recognised nor accounted for Motutapu's culturally historic landscapes. The plans generated a severe backlash from archaeologists and cultural heritage advocates who were outraged that anyone would want to revegetate over one of the most complex and intact Māori landscapes in the Auckland region.

In this context, it is noteworthy that over the past two decades there has been marked conflict and heated debate between those supporting ecological restoration and those wanting to preserve the cultural heritage of the island. For example, in late 1991 three members of the University of Auckland's Centre for Conservation Biology produced a plan to ecologically restore Motutapu, and their work was published three years later (Miller *et al.*, 1994). They believed a program of pest eradication and revegetation of the island would create the best example of an island open sanctuary in the world, especially when coupled with open public access and interpretation. The plan encapsulated the attitude that New Zealand's natural heritage required protection and that islands were ideal places to achieve such ends. The plan also drew on nationalist sentiments; the authors calling for New Zealand to maintain its status as a nation with innovative and ambitious conservation practices, and take up this conservation vision for Motutapu in order to protect and conserve the native habitats and threatened species of New Zealand. The plan was to plant two-thirds of Motutapu in native vegetation and gradually (re)introduce native species, and particularly birds. In the wake of this proposal being made public, a number of volunteer tree-planting projects were established, and a plant nursery was opened in March 1992 by the Duke of Edinburgh, Prince Philip, then the International President of the World Wide Fund for Nature (Department of Conservation, 1992). While the tree-planting program was being publically applauded, the plan raised alarm in archaeological circles. A month after Prince Philip's visit, the revegetation of Motutapu was discussed at the annual New Zealand Archaeological Association conference. Several key concerns were expressed, among them that archaeologists had not been consulted, and that the significant archaeological landscape of the island could be destroyed or obscured as a result of tree-planting. Members of the conference sent a joint letter to both the Department of Conservation and the World Wildlife Fund for Nature expressing their collective apprehension about the plan's effects (Gardener, 1993).

Significant debate continued in the mid-1990s. In June 1992, the Department of Conservation released a working plan for the revegetation of Motutapu based on the plan from the University of Auckland's Centre for Conservation Biology. The plan provoked numerous submissions to the Department from members of the public and cultural heritage managers (and particularly from archaeologists). Their key concern was that the cultural heritage values of the island should be given higher priority (Dodd, 2007). Irwin *et al.* (1996, p. 254) explained that the plan erroneously 'maximised natural history values, of the island, of which little remained, and minimised the cultural heritage values of the island, of which much remained'.

Having analysed the content and significance of these submissions, the Auckland Conservation Board decided to convene a public workshop involving the Department of Conservation and all stakeholders (Gardener, 1993). John Craig spoke of the original plan put forward in 1991 and explained that the goal was to create a functioning ecosystem and not simply an educational showpiece. Prominent New Zealand archaeologist Janet Davidson explained that the revegetation of the island would not restore a past landscape but rather would alter an important cultural landscape. Meeting facilitators organised those in attendance into small discussion groups, each to formulate working plan objectives to present in plenary. Gardener (1993, p. 37) wrote that the results from one such collective, group five, caught everyone's attention: the goal the group conceived was to 'protect and enhance most archaeological sites while providing sufficient habitat to accommodate threatened species'. The island would therefore feature cultural heritage *complemented by* native fauna and flora. This goal was adopted by the meeting, and included a pest and weed eradication program and an invitation to Māori tribes (iwi) to participate in the plan – the last of these highly significant on the basis that, from the time that European settlers had purchased the island, the iwi associated with the island had been alienated from the land. In this sense, the plan advanced by group five provided for a measure of living cultural heritage restoration.

In 1994, and on the basis of the aforementioned multilateral negotiations, the Motutapu Restoration Trust was established to restore and enhance the natural *and* cultural heritage of Motutapu. The two principal goals of the Trust are to restore the island's 'natural landscape [in ways] similar to that which existed on Motutapu after the Rangitoto eruption around 600 years ago [and to value the] cultural landscape handed down by Māori, early settlers, farming and the military' (Motutapu Restoration Trust, n.d., page unknown). Demonstrably, there was concerted effort to enhance, protect, and preserve *both* the natural and cultural heritage of the island. However, translating this vision into actuality has been difficult because of the entrenched values of different groups (Bade, 2010).

Since the establishment of the Motutapu Restoration Trust, approximately 100 hectares (247 acres) of the island has been planted by volunteers and numerous native birds have been (re)introduced (Figure 6.3). A considerable

Figure 6.3 The first volunteer-planted forest of Motutapu, located at the Home Bay
       Valley. Photograph by David Bade.

amount of research goes into selecting species to plant on the island which
may have been there before human settlement. Such selection is informed by
environmental data from past archaeological investigations and by sourcing
seed from the local area (Dodd, 2007). These procedures ensure the planting
is as 'natural' or 'authentic' as possible. When planting, volunteers are also
directed to mix the types of trees they are planting 'as nature would do'.

As on Rangitoto, on Motutapu there remain conflicting views regarding
the status and importance of cultural heritage on the island. In island restoration contexts, historic sites are often viewed as obstacles or impediments to
the goal of ecological restoration, and archaeologists and cultural heritage
managers are often in conflict with environmentalists and ecologists.

Conclusion

The foregoing discussions about Rangitoto and Motutapu clearly suggest the
desire among New Zealanders to 'restore' islands back to a 'natural' state.
This social impetus to do something 'for the environment' has increased in
popularity from the 1990s and relates directly to New Zealand's image as
clean, green, and pure. As highlighted by the thousands of volunteers who

have now planted trees on various Hauraki Gulf islands, restoration projects, reforestation, and ecological restoration have become popular recreational – and indeed romanticised – activities for Aucklanders. In light of the foregoing, it seems reasonable to argue that the nature conservation ethic has entrenched itself in the New Zealand psyche, and to suggest that it has led to islands being reserved for their natural values and ecologically restored in ways that result either in negative outcomes for the cultural heritage of the islands or in signals that such heritage and the place values that attend it are presently under-valued.

In New Zealand, a strong and active nature conservation ethic has developed since the mid-nineteenth century as nostalgia about the colony's, and then the nation's, forests and fauna grew in response to dramatic landscape transformations from forest to pasture and in light of expressed and explicit concern about shrinking levels of native biodiversity and the loss of landscapes. Māori settled among these islands only 800 or so years ago, and the dominant Pākehā population settled less than 200 years ago and thus, because of New Zealand's comparatively recent history, this place has become internationally renowned for having spectacular and significant natural heritage features of significant longevity. As a result, in New Zealand there has been an increasing desire to protect nature by means of the designation of reserves, and to 'restore' landscapes, and particularly islands, back to their 'original' states.

Islands are conducive to the constitution of particular values that serve to prioritise nature preservation. Likewise, the urge to excise and elide exotic species and evidence of past human activities is one readily indulged on islands, the boundaries of which render ensuing conservation achievements more palpable. As a result, although the hundreds of years of Māori and European influence on the land have left their mark in every corner of New Zealand, the nation's island reserves are predominantly valued for their perceived naturalness. Occasional items or sites of cultural heritage such as lighthouses or historical cottages may be considered relatively significant but, more often than not, New Zealand's island reserves continue to be the repositories and locations of past human endeavours considered too slight in importance to be conserved in the face of natural heritage conservation; at risk in the process are histories of farming, abandoned settlements, sealing, whaling, quarantine stations, and World War II posts. It may now be timely to reconsider such trends and revalue the remarkable cultural heritage that exists on New Zealand's offshore islands.

Note

1 These island wildlife reserves are considered the first of their kind in the world – that is, areas of land legally protected by Western law (Diamond, 1990; Star, 2014; Star and Lochhead, 2002; Young, 2004). However, it is acknowledged that people have been protecting wildlife on and off islands for millennia using traditional practices such as in the protected ancient forests of India, China, and Nepal, or in the sacred groves of Western Africa (Holdgate, 2013, p. 2).

7 'The good garbage'

Waste-to-energy applications and issues in the insular Caribbean

Russell Fielding

Introduction

Throughout the world, small islands not adjacent to larger landmasses experience several common problems related to infrastructure and public services. Aspects of the physical geographies of these islands – especially their insularity and remoteness – require that various infrastructural systems be designed to operate independently of larger grids and to fit the small scale. Among many challenges on such islands are those related to energy production, waste disposal, and water supply. Three related observations arise from knowing that such is the case.

First, according to the International Scientific Council for Island Development (ISCID), small islands are ideal sites for the testing and refining of sustainable energy production systems:

> traditional limitations in the energy field, [such as] distance from the major grids, small scale, distribution difficulties and the lack of large conventional markets, are more than offset by the extreme abundance of renewable energy sources ... in island regions. In fact, we would go as far as to say that islands have become genuine laboratories of the future of energy sustainability.
>
> (Marín, 2004, p. 2)

Yet, while the ISCID's optimism is well-received in academic settings where the idea of the island as laboratory has gained purchase since at least the earliest days of biogeographical study (Sauer, 1969), those involved in many island governments and energy industries remain unconvinced. As noted by Notton *et al.* (2011, p. 652), 'the most usable power plant for small islands is diesel engines'. This trend may be on the cusp of changing, however, as fossil fuel costs continue to increase, and as both islanders and tourists demand more sustainable solutions.

Second, an island's capacity to handle its municipal and industrial waste is directly related to the size of its physical land area and to its population. On islands with extreme population density, such as Manhattan, the export of garbage seems the only option. People living on larger or more sparsely populated islands may relegate some of their land area to landfills. Incineration is also widely practised – both as centralised and household activities. Issues of air pollution are well-documented with regard to incineration. One major argument against the implementation of sustainable waste management solutions has been that landfills are relatively cheap and abundant in mainland settings and on large islands such as Britain (Read *et al.*, 1998). However, this line of reasoning applies less in small island contexts. By virtue of islands' naturally limited land areas, islanders have added incentive to develop efficient methods of waste disposal.

Third, those living on small oceanic islands – especially those without significant surface water or groundwater reserves – can experience the plight of Coleridge's *Ancient Mariner*: 'Water, water everywhere / Nor any drop to drink.' People living on small, dry islands often rely on rainwater catchment as their primary source of fresh water, which leaves little recourse during droughts and regular dry seasons – except for rationing, doing without, or importing fresh water. Climate change further exacerbates the uncertainty of supply. Desalination is an effective option on some islands, but many more are unable to provide enough water through this process, owing to the inherent expense – both in terms of finances and energy consumption. Yet, as Swyngedouw (2013) has noted, varied politico-social issues affect the development of desalination facilities in mainland settings and there is no reason that one should not apply – and even amplify – his findings in island contexts.

Several island communities throughout the world have endeavoured to resolve these challenges by means of technological development and investment. The key to such solutions is often found in combining efforts and integrating technologies in order to solve multiple infrastructural goals at once. Waste-to-energy facilities are one salient example of such integrated technologies on small islands (Díaz, 2011). While many varieties of facility exist, most involve the capture and redirection of thermal energy released from incinerating municipal and/or industrial waste. This energy is then used to produce electricity or desalinise seawater, with scrubbers removing pollutants from the exhaust smoke of waste-to-energy facilities, albeit with varying degrees of success. Owing to their efficiency and capacity to address several sustainability challenges, then, waste-to-energy facilities are often seen as ideal for small island settings. Notwithstanding, these facilities are expensive and require large initial investments.

Wealthy island populations, or those with political ties to others residing in wealthy nations, are often in the best position to invest in waste-to-energy facilities and other sustainable technologies. Such is the case in St Barthélemy but not its 'neighbour', St Croix (Figure 7.1). As an overseas collectivity (*collectivité d'outre-mer*) of France, St Barthélemy benefits from cultural, political, and economic ties with France. Additionally, the island's niche focus on luxury tourism brings foreign capital into St Barthélemy at a pace unrivalled by most of the island's Caribbean neighbours. Indeed, the local government

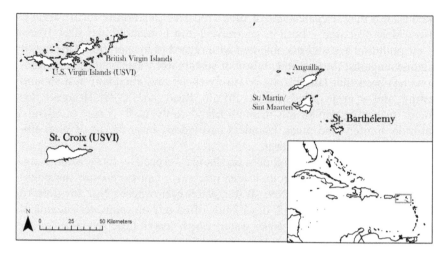

Figure 7.1 Locations of St Croix, St Barthélemy and surrounding islands. Cartography by Benjamin McKenzie. Reproduced with permission.

of St Barthélemy has recognised its peculiar position within the region and sought to capitalise on the opportunities presented to serve as an example of sustainable development to the rest of the insular Caribbean, as evidenced by the government's investment in waste-to-energy facilities. Currently, a combined waste-to-energy facility operates outside of St Barthélemy's capital, Gustavia, providing thermal energy to the island's seawater desalination plant and offsetting that normally energy-intensive industry's electricity demand. By comparison, on St Croix – one of the United States' Virgin Islands and an unincorporated territory of the United States – people have witnessed the failure of a proposed waste-to-energy facility and continue to struggle with issues of waste management and energy production.

This chapter investigates these two cases: one of measured – yet challenged – success, and one – at least thus far – of failure. That investigation is achieved first by examining the case history of the St Barthélemy waste-to-energy facility; then by considering the case of St Croix; and then by concluding with a comparison of the two islands: their geographies, current status, and outlook with regard to waste-to-energy technologies.

St Barthélemy

Located in the Leeward Islands of the Lesser Antilles, St Barthélemy (also called St Barth or St Bart's) is a small island of about 23 square kilometres (8.8 square miles) (17.9° N 62.8° W). It was sighted by Columbus in 1493 on his second voyage and named for the navigator's brother, Bartolomeo. Prior to European discovery, St Barth had been known as *Ouanalao* by the Carib people who, owing to the island's lack of fresh water, visited occasionally but made no permanent settlements. In retrospect, the lack of permanent indigenous settlement should have foreshadowed problems for the coming colonialists.

The island was colonised by the French in 1648, ceded to Sweden in 1784 in exchange for free trading rights in the port of Göteborg, and kept as Sweden's only Caribbean territory until 1878, when it was returned to France. In St Barth today, one sees multiple references to the near-century of Swedish association, including the blue flag with a yellow Scandinavian cross, which flies from many of the island's flagpoles. Printed text in some public and private establishments is translated into both English and Swedish, as a nod to the island's history; many street signs in Gustavia present the French name as well as the Swedish; and Gustavia is paired with its sister city Piteå in Sweden.

St Barth remained a poor colony for the first half of the twentieth century, isolated without an airport until the late Rémy de Haenen – one of the island's most celebrated residents – cleared an airstrip that today remains one of the shortest and most difficult landings in the world. During the 1950s, St Barth was 'discovered' by American and European celebrities and millionaires. Some wealthy families established their presence more permanently: the Rockefellers built a large house on the island's west end, near Colombier, and the Rothschilds did the same on the east coast, by Grand Cul-de-Sac. Infrastructure provision began to catch up: electricity came in 1962 during de Haenen's tenure as the island's mayor. However, fresh water remained a problem. Each home relied on its own cistern and on two larger reservoirs on high points at either end of the island. During ensuing decades, more development gradually transformed St Barth from a small dry island known for duty-free port, fishing, and salt ponds, to a small dry island known for luxury tourism.

Consume and discard

With a population in 2011 of 9,057 (Cotis, 2011), and visitors numbering in the tens of thousands annually, the infrastructure demands on St Barth are considerable, especially during the winter high season for tourism. Then, cars and motor scooters choke the island's steep and narrow streets, and parking spaces along the boardwalk in Gustavia become as sought-after as the quayside berths where multi-million dollar yachts are docked.

Owing to its negligible agricultural and manufacturing outputs, St Barth imports nearly everything that residents and visitors consume. The luxury niche market targeted by those in the tourism sector has led to what Cousin and Chauvin (2013, p. 191) identify as 'competitive consumption'. They also describe late-night revelries in which business magnates, celebrities, and other members of the super-rich class attempt to out-spend and out-consume one another, rewarded with exclusive (yet highly visible) seating in the island's gathering places where they may consume imported beverages costing thousands of dollars per bottle.

Packaging and detritus associated with these commodities are collected at the island's waste disposal plant, located just outside Gustavia in what some maps identify as the *Zone Industrielle*. Fronted by a village called Public, this area is home to the commercial port, where vessels not bearing tourists load and unload their wares. Here, enormous piles of solid waste are sorted into three categories: that which can be burned, that which must be sent off-island, and that which can be repurposed. This industrial zone features none of the sights and smells of the rest of the island. Decomposing organic garbage replaces the scent of frangipani, bougainvillea, and fresh-baked *pains au chocolat*, while separated piles of crushed glass, cubed aluminium, non-working appliances, and general garbage stand in stark relief to the raked beach sand, quaint villas, and landscaped gardens of the St Barth that is seen by most visitors. Although the site remains largely unseen, residents and visitors alike experiences its effects. Each day, a careful process unfolds in which industrial and municipal solid waste is brought to the facility, sorted into categories, and processed. In theory, sorting begins in the home or at the public garbage receptacle. An island-wide information campaign includes the provision of instructional fliers to residents and placards that urge tourists, in English, to put waste '*in the good garbage*' in order to facilitate the sorting process that happens at the incinerator site.

When the waste is sorted, the first category includes items that can be recycled off the island. At the incinerator site, in the spaces allotted for the appliances, batteries, aluminium cans, and steel, stand neat stacks of similar things, all sorted under signs bearing their designations – here only in French, as there is no need to translate instructions into English and certainly not into Swedish. Vegetable-based oils and petroleum products stand in separate vats. All of this waste will be shipped to various ports in other places – Guadeloupe, Miami, France – where it will be recycled or turned into scrap.

The second category is made up entirely of glass, which can be repurposed locally on St Barth. After being separated by hand from all other forms of waste, glass is crushed into a fine powder for industrial use. Those whose job it is to sort the glass spend their shifts – heavily gloved – pulling corks from empty wine bottles, twisting off caps from expensive *eaux minérales*, and shaking burned-out sparklers from expensive bottles of champagne (Cousin and Chauvin, 2013). This glass powder will later provide insulation for water pipes and electrical conduit running under the island's roads and sidewalks.

The third category is combustible waste. All vegetal trimmings, paper products, and other organics, as well as most plastics are piled as fuel for the island's incinerator. One day of drying under the tropics sun is usually sufficient preparation for the combustibles. Once dried, the fuel is lifted into a chute where it feeds a constantly burning flame. The incinerator needs at least 25 metric tons of material per day to maintain its optimal burn rate: 35 is better and 50 is the maximum. Just enough material is produced to make incineration sensible. During the slow season of summer, when jet-setting tourists are more likely to be found in the Mediterranean than the Caribbean, the incinerator occasionally shuts down, owing to a lack of fuel. This seasonality reflects the absolute dependence of St Barth on its wealthy tourist clientele. Without them, or the waste they generate, the incinerator at the waste-to-energy facility stops burning. When the flame goes out, it can take days to return the operation to its optimal temperature.

A closed-loop system for evaporation and condensation of water is connected to the incinerator. Once water is heated by the combustion of waste, steam is sent to the nearby desalination plant, where its thermal energy is used to power the production of fresh water. During the summer low season, the steam alone provides enough energy for the production of fresh water through evaporation. When demand is high, an electricity-powered reverse osmosis process is added. Imported diesel serves as the fuel to generate electricity.

Critics of waste incineration often focus on the air pollution inherent in the process. The incinerator on St Barth certainly introduces chemicals and particulates to the atmosphere, although scrubbing processes are in place to limit both. Nevertheless, more independent research is needed to quantify the concentrations of these pollutants; this gap represents a remarkable opportunity for an atmospheric chemist to conduct serious and necessary research in a breathtakingly beautiful setting. However, when evaluating the costs and benefits of a system such as that on the island, it is important to consider the processes that are being replaced. In the case of waste management on St Barth, the incinerator largely replaces two methods of waste disposal once common on the island. According to long-time residents Alexandra Deffontis and Bruno Magras, most residents either burned household waste at home or simply dumped it directly into the sea. Each of these people, participants in my research, now plays a direct role in the waste-to-energy program on St Barth.

In another conversation with me, Magras – a St Barth native and the island's political leader (*Président de la Collectivité d'Outre-Mer de Saint-Barthélemy*) – has reflected on the gravity of his position: 'I'm concerned about my island, my future, my kids' future. I'm not out to destroy what I received' (personal communication, 19 March 2013). While the building of the current waste-to-energy incinerator in 2002 was based on a decision by the French government, Magras takes credit for the island's original municipal incinerator, built in 1979. This first incinerator did not produce electricity or energy for desalination, but it did serve to centralise waste disposal, and offered an alternative to the then-current practices of household-scale garbage burning and nearshore dumping. Magras's approach to renewable energy production is nuanced, however. He is against wind power – the large, offshore turbines would be 'too ugly'; against large-scale solar energy – open land is too scarce; and against a proposed undersea cable supplying power from nearby St Martin – a plan that would increase the island's dependence as well as its vulnerability to hurricanes. Magras does support the capture of solar energy at the household level; however he worries that installation of photovoltaic cells atop the famous red tile roofs of Gustavia would reduce a popular aesthetic. Magras's dedication to the (perhaps illusory) notion of an 'unspoilt' island environment that turns him away from wind and solar development has not dissuaded him from the development of waste-to-energy technology, perhaps by virtue of its 'hidden' nature, tucked away in a corner of the island usually unseen by tourists.

Deffontis, another St Barth native, now works as an official at the incinerator (her title, *Directeur du Service de Propreté*), and she praises its effectiveness in dealing with the 'biggest problem' faced by the island during the high tourist season: the availability of fresh water. According to Deffontis (personal communication, 20 March 2013), hotels and rental villas are given priority in the distribution of municipal water. Prior to desalination, when rainwater catchment was the primary source of the island's fresh water, households would carefully measure the reserves in their dwindling private cisterns as wealthy tourists lounged by 'infinity pools' and deckhands washed their employers' yachts along the quay. Today, water is more readily available, but only inasmuch as combustible garbage is produced on the island. As examined below, several challenges threaten to disrupt this precarious system.

Challenges to waste-to-energy effectiveness in St Barthélemy

While the incinerator does appear to have achieved a form of equilibrium in St Barth, the system is not without challenges. Here, I discuss three issues related to the effectiveness of the incinerator that are most commonly mentioned on the island.

Hurricanes

St Barth is often in the path of Atlantic tropical cyclones. In recent times, major storms to make landfall on St Barth include Hurricane Earl in August 2010, and Hurricane Gonzalo in October 2014. The French authorities on St Barth use the colour-based French tropical cyclone scale as opposed to the Saffir-Simpson Scale used throughout much of the English-speaking Caribbean and North America. This *Tableau des Alertes Cycloniques* ranks tropical cyclones by colour, based on average windspeed over a ten-minute interval, and was developed by Météo-France (2013), originally for use in the southern Indian Ocean to monitor storms near La Reunion. The incinerator is able to maintain operation during less-powerful storms (classified as yellow and orange). It must stop during larger hurricanes (red).

Lack of public compliance with waste management procedures

When the incinerator began operation in 2001, islanders were asked to sort their garbage at home. Prior to that time, no sorting had been necessary. Some residents refused to participate when the sorting regimen began – and in some cases maintain this resistance to this day. Garbage collectors would find metals and plastics in bins meant to contain only organic waste or they would find all of a household's waste combined in the same receptacle. These small acts of civil disobedience were interpreted as statements against the regulation of waste disposal and the continued modernisation of the island. They may even represent the dissenting voices of islanders who are not completely in favour of the accommodations being made for a growing population, driven mainly by the continual increase of the tourism sector. However, the lack of compliance continually puzzles the incinerator's managers, especially considering the multiple grassroots environmental protection campaigns created on in St Barth, which range from one promoting homemade cigarette disposal stations at many of the island's beaches to others championed by *St Barth Essentiel*, an organisation started in 2009 to preserve natural and cultural heritage.

The government responded to the public's failure to thoroughly sort household waste by streamlining the waste management system and marketing it using bilingual flyers and placards. The campaign features a stylised pelican – reminiscent of the birds on either side of the official seal of St Barthélemy – reminding residents that they need to manage 'ONLY 2 GARBAGE BAGS!' ('*2 POUBELLES SUFFISENT!*'). One bag is for combustibles, the other for recyclables. The campaign met with limited success but its advocates persist in seeking to raise awareness about the benefits of waste management. Their efforts include placing receptacles in visible public areas; distributing instructional, bilingual literature that stresses the ease and importance of separating household garbage; and publishing news articles urging compliance (see Anonymous, 2009). While some parallel English slogans are found on the literature, only the French side proclaims that '*Trier c'est Gagner!! S'abstenir c'est detruire!!*' ('Sorting is winning!! Refraining is destroying!!').

Composting

The government on St Barth has been exploring the possibility of introducing an island-wide composting program. While composting is usually seen as an environmentally beneficial activity, incinerator managers worry both that the program will divert combustible organic material away from the incinerator, and that they may fail to receive enough fuel to keep the operation going efficiently. At time of writing, for want of fuel, incineration is occasionally halted during the slow summer season. The establishment of a new waste management stream may divert enough material away from the incinerator to make its operation inefficient at best, and impossible at worst.

Further, it is unclear what the ultimate use and usefulness of composting will be given the limited presence of agriculture on the island – beyond, perhaps, the occasional household garden. Certainly, the kitchen garden has long provided household self-sufficiency in the Caribbean (Fielding and Mathewson, 2012; Kimber, 1966; Richardson, 1983). However, with its dry climate and rocky soils, St Barth has not supported small-scale agriculture at a level comparable to some of its Caribbean neighbours. Still, in March 2013 the government acquired a piece of land adjacent to the incinerator to be used for composting. The site has since been relocated but plans to host a municipal composting operation remain. When the composting program will begin, how it will be received, and what its results will be remain unknown. The threat that the scheme poses to the supply of fuel for the waste-to-energy facility underscores the island's tenuous reliance on imported goods. It also raises the possibility of importing waste destined specifically for the incinerator to overcome future shortages, due either to competition for the waste itself or to seasonal slowdowns in tourism sector. While, to my knowledge, St Barth has not considered such options, it is a strategy that has worked in other locations such as Sweden (Olofsson *et al.*, 2005).

Application to St Croix of lessons learned in St Barthélemy

St Barthélemy is by no means unique as an island that has a complicated relationship to waste-to-energy technology. Throughout the Caribbean, one finds islands where governments have considered, built, or started to build such facilities. One island where waste-to-energy technology has been discussed but not implemented is St Croix (17.7° N, 64.7° W).

Sustainability and bird strikes

Currently, the majority of the waste produced on St Croix is dumped in the Anguilla Landfill, a large, open, unlined dump, located on a small peninsula on the island's south coast, adjacent to the Henry E. Rohlsen Airport. Note that, despite its name, the landfill is located on the island of St Croix, not on the nearby island called Anguilla. This facility is problematic owing to the large flocks of birds that the landfill attracts. For aircraft taking off to the east or approaching from the east to land, flocks of birds can be present in the flight path. Produced by Jeppesen Incorporated, and known by pilots as a 'Jepp chart', an information diagram for this airport includes a bold warning near the top of the page: 'Birds in vicinity of airport.' The chart also includes a displaced landing threshold for aircraft approaching from the east, where the landfill is located, to keep them at higher altitude until past the landfill (Figure 7.2). According to the Wildlife Strike Database maintained by the United States Federal Aviation Administration, there have been 46 documented bird strikes involving aircraft taking off or landing at this airport since 1992. The reporting of bird strikes is voluntary and as such, this figure probably represents less than the total number of incidents that have occurred in St Croix.

Aware that this problem will likely increase as the Anguilla landfill expands in size and use, the Federal Aviation Administration has toughened its stance in relation to St Croix. According to Collins (2003), and corroborated by reporters employed by other news outlets, owing to the hazard presented by

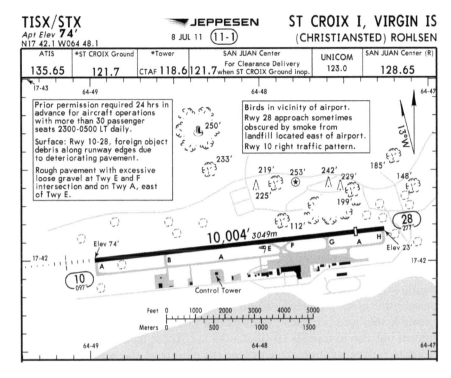

Figure 7.2 A section of the 'Jepp chart' showing the Henry E. Rohlsen Airport in St Croix, USVI. Note the boxed warning about 'birds in vicinity of airport' as well as the displaced landing threshold represented by a white bar toward the east end of the runway. Image reproduced with permission of Jeppesen Sanderson, Inc. Reduced for illustrative purposes only. NOT FOR NAVIGATIONAL USE. Jeppesen Sanderson, Inc. Copyright © 2015.

the accumulation of birds near the runway's east end, the Federal Aviation Administration has suspended its annual grants to the Virgin Islands Port Authority, the 'semi-autonomous agency that owns and manages the two airports and the majority of the public seaports in the [United States Virgin Islands]' (Virgin Islands Port Authority, 2013, n.p.). This suspension of grants is intended to last until the Anguilla landfill is closed and the bird problem is mitigated. Without the funds supplied by these annual grants, it is doubtful that the Virgin Islands Port Authority will be able to properly maintain the physical plant of the Henry E. Rohlsen Airport to the standard required by the Federal Aviation Administration for airports handling commercial flights. In this case, commercial flights would be suspended and passengers would only be able to arrive in St Croix by sea or by private aircraft; undoubtedly such circumstances would have an enormously detrimental impact on tourism in St Croix. According to the United States Virgin Islands Bureau of Economic Research (2013), 132,958 tourists arrived in St Croix by air in 2013. Tourism is the major industry in the United States Virgin Islands, and has become increasingly important on St Croix since the 2012 closure of the Hovensa oil refinery, once the largest private employer in the Islands. Any potential disruption to air traffic into St Croix would further affect the island's already-weakened economy.

In addition to the risk presented by birds in the vicinity of the airport, the Anguilla landfill is reaching capacity. Problems related to the volume of garbage at the landfill include fires and collapse of enormous garbage piles – each a significant hazard in its own right, with the former producing smoke that presents further risk to aviation.

There are strong incentives to remediate the problems presented by the Anguilla landfill that include, but go beyond, aviation safety. Given its limited land area St Croix faces infrastructural challenges similar to those experienced on St Barth. Populations on both islands must constantly make wise choices in land-use planning. Relegating an expanding portion of the limited space available to a landfill is not a sustainable solution. Further, as infrastructure demands continue to increase, especially with regard to fossil fuel consumption in the production of electricity for industrial and municipal use, renewable solutions become more attractive to the local population and government but also to tourists visiting the island. Those on St Croix have good reason to pursue solutions that include the closure of the Anguilla landfill, based on the Federal Aviation Administration mandate as well as broader principles of sustainability and good stewardship of the limited land area of this small island.

The reason for the delay in closing the Anguilla landfill seems to be that the island government has not implemented any alternative means of dealing with its waste. Currently, waste is baled – crushed, compacted, and bound – before being deposited in the landfill, a process that reduces the accessibility of the garbage to birds and maximizes the use of space but does not fully alleviate the problem. Rather than closing and covering the landfill – the Federal Aviation Administration's recommended course of action – the Waste Management Authority of the Virgin Islands (the Authority) has recently petitioned the territorial government for permission to expand its area, allowing continued usage of the landfill for the next three to seven years. Even with the conversion from an open dump to a repository for baled garbage, the problems of birds in the vicinity of the airport and an increasing area on a small island being given over to waste-holding have not been fully remediated.

An attempted solution

St Croix is in need of an alternative, sustainable method of handling its waste – because of the hazards to aviation, which are significant, and to maintain the environmental integrity of the island, which is expected by tourists and residents alike; the territorial government of the Virgin Islands knows this and has been actively seeking a solution. In August 2009, after exploring multiple possibilities – including shipping garbage off-island, the Authority contracted Denver-based Alpine Energy Group, LLC to construct a waste-to-energy facility on St Croix. Just as on St Barth, waste-to-energy technology has been seen as an attractive solution to St Croix's dual problems of waste disposal and energy production. On St Croix, waste-to-energy would have the added benefit of improving the safety of air travel as well. The Authority apparently believed that the project would be successful, as it was willing to commit to the project with an initial investment that exceeded US$10 million. Alpine Energy made a similarly substantial investment. Clearly both parties expected the project to succeed.

Yet, amid controversy, the waste-to-energy proposal was abandoned in 2012. The narrative of the failure of the St Croix proposal is one steeped in controversy. Local newspaper articles document the transition from hopefulness to doubt to mistrust to animosity to defeat. A sample of local headlines includes the following:

2009

11 August	*US Virgin Islands Government Signs Contracts With Alpine Energy Group. To Develop Territory's First Alternative Energy Plant*

2010

12 January	*Alpine Energy Officials Answer Public's Concerns Monday Night*
8 March	*Senators Garbage Alpine Energy Deal*
22 October	*Alpine Energy nears close on St Croix WTE*

2011

27 April	*Alpine Withdraws Application to Build St Croix Waste-to-Energy Plant*
20 August	*Will It Be Alpine Energy After All?*
2 December	*Alpine Energy Group Request Needs Greater Scrutiny*
9 December	*Alpine's Project Again Rears its Ugly Head*
23 December	*Can a Waste-to-Energy Plant be Environmentally Safe?*
9 February	*Senate Rejects Alpine Lease, Killing Garbage-To-Energy Plan*

As these select headlines indicate, when talk of the planned waste-to-energy facility began it was viewed as a positive move toward sustainability and greater aviation safety, agreed upon by the St Croix government and the contracting company. However, this outlook deteriorated over the course of about two-and-a-half years. From the perspective of Don Hurd, president of Alpine Energy (personal communication, 24 October 2012), the proposal failed because of a lack of communication and cultural understanding between his Colorado-based corporation and the public on St Croix. Local newspaper stories seem to corroborate this assessment, emphasising too the effects of mistrust inherent in geopolitical relationships that have characterised colonial histories such as that typifying the United States Virgin Islands.

Of St Croix's colonial history ... and the problems that remain

Variously claimed and settled by the Spanish, English, Dutch, and French, St Croix's colonial economy was tied to monocrop agriculture – primarily sugarcane – but also cotton and tobacco, and it was reliant upon slave labour instigated under Danish colonial rule, which followed the sale of the island by the French in 1733. The islands were then sold again to the United States in 1916 as part of the Treaty of the Danish West Indies.

The present-day United States Virgin Islands – St Thomas, St John, St Croix, and many smaller islets – now comprise an organised, unincorporated United States territory. Residents are United States citizens but cannot vote in federal elections and do not pay United States federal income tax. The economy of the islands is heavily dependent upon the United States, with tourism as the major industry. Petroleum had been a close second until the recent closure of the Hovensa refinery. As in many colonial or postcolonial geopolitical environments, there are significant levels of mutual distrust between the colonised and the colonisers (de Albuquerque and McElroy, 1999; Navarro, 2010). This distrust is evident in the feelings of local residents toward 'outsider scientists' (Grace-McCaskey, 2012, p. 85) and others attempting to establish changes for the sake of environmental protection and sustainability.

Despite the cross-cultural communication difficulties inherent in an offshore dependent territory with such a multinational colonial history, the majority of stakeholders and residents on St Croix agree that something needs to be done about the Anguilla landfill. Besides the aviation risk presented by the large flocks of birds that the garbage attracts, the landfill has regularly been catching fire for over a decade (Anonymous, 2012; United States EPA, 2001). According to the United States Environmental Protection Agency, it is likely that the fire outbreaks actually stem from slow-burning subterranean fires that have been smouldering constantly for many years (United States EPA, 2001). These fires, with their lack of monitoring and control, emit unknown quantities of hazardous fumes and particulates into the atmosphere. The smoke from the fires presents its own hazard to aviation, as mentioned above. Additionally, rainwater that percolates through the landfill has the potential to leach hazardous chemicals into the groundwater and eventually the ocean, as the landfill is not lined with a waterproof or filtration layer. Thus, while a waste-to-energy facility would certainly introduce some level of pollution to the atmosphere and ocean, it would likely be orders of magnitude less than what is currently entering the environment through the uncontrolled burning and leaching at the Anguilla landfill. The waste that the facility would convert to electricity is already burning. A waste-to-energy-enabled incinerator would simply burn the waste in a controlled environment, where it would be converted to electricity, and the exhaust would be treated before release into the atmosphere. Unlike St Barth, St Croix has a ready repository of waste to be burned in addition to the waste its communities produce daily. Alpine Energy did not include 'garbage mining' as part of its waste-to-energy plan, but this is not to say that it could not be included in the design of a future system.

Both the Federal Aviation Administration and the Environment Protection Agency have issued clear directives that economic consequences, require that the Anguilla landfill be closed, and refuse to sanction a waste-to-energy facility. At present, St Croix's lack of compliance with those directives can be traced directly to a lack of alternative methods to handle the island's municipal and industrial solid waste. Meanwhile, the closure of the Hovensa petroleum refinery took away not only jobs from St Croix, but also the island's major source of fuel oil for its electricity generation plants. St Croix now imports fuel oil for its electricity generation (United States Energy Information Administration, 2013). For energy production on St Croix, the government of the United States Virgin Islands has then considered solar (Lewin, 2012), wind (United States Energy Information Administration, 2013), and biomass (Viaspace, Inc., 2013). Although each is feasible, none has the potential to alleviate the need to keep the Anguilla landfill open, nor to address the aviation hazards that the landfill presents. From the perspective of Alpine Energy, the fact that waste-to-energy and other methods of renewable energy generation have not been implemented on St Croix is due primarily to a failure to communicate effectively across cultures (Hurd, personal communication, 24 October 2013). Based only on this explanation, it follows that the successful waste-to-energy implementation on St Barth is due to better cross-cultural communication on that island. While communication related to the waste-to-energy facility development has been suboptimal on St Croix, the communication breakdown itself may stem from broader, historical differences in the islands' colonial relations with their respective mainlands and the continuation of those relationships into today's tourism-based economies, which are being enacted against the backdrop of the current postcolonial paradigm.

Discussion and conclusions – islands of differences

One key question that illuminates the complexity of the geographies of St Barth and St Croix remains unanswered: can a facility similar to the waste-to-energy facility on St Barth be established on St Croix to address that island's waste management and energy production needs, while providing relief to those pilots and air passengers who worry about bird strikes and flight paths obscured by smoke? On the surface there appear to be similarities between the two islands that would indicate a similar outcome: both are small, subnational islands in the Lesser Antilles. Each has a past touched by slavery, was associated with the French and with a Scandinavian power, and now each is connected to a major Western power. Both islands experience a dearth of locally available natural resources and have turned to imports to meet most of their material needs. Both island economies heavily depend upon tourism. Arriving at each island by air requires a tricky landing.

However, there are some key differences, historically, economically, and culturally, that may have favoured the difference in outcome that exist between St Barth and St Croix. Despite the superficially similar French-Scandinavian histories, the economic, cultural, and political trajectories of St Barthélemy and St Croix diverge in important ways. St Croix was host to plantation slavery under the Danish regime and to a lesser extent under the French (Dookhan, 1994). While the Danes were the first Europeans to abolish the trans-Atlantic slave trade, slave ownership was not immediately forbidden in the colonies. Loftsdóttir and Palsson (2013, p. 45) presciently point out that, following the abolition of the Danish slave trade in 1802, although 'property in people was no longer legal in Denmark, different principles applied in the colonies. Freedom, human rights, and dignity had their own geography.' Today, according to the United States Census Bureau, 76 per cent of the population of the United States Virgin Islands identifies as Black or African American. The majority of the island's current inhabitants can trace their ancestry to the enslaved African labourers brought to St Croix by European colonialists.

In contrast, while slavery was practised on St Barth, there were no large, agricultural plantations. As such, it was primarily smallholder, subsistence agriculturalists who dominated the institution. The practices of these slaveholders gave St Barth the reputation, according to the eighteenth-century anti-colonialist writer Abbé Guillaume-Thomas Raynal, as 'the only one of the European colonies established in the new world where free white persons deign to share agricultural tasks with their slaves and to labor in the field alongside their subordinates' (Robequain, cited and translated in Lavoie *et al.*, 1995, p. 380).

To be sure, slavery on St Barth – as with everywhere it was implemented – was a degrading and deplorable practice. Its offence gave rise to at least one local slave rebellion, in 1736, which resulted in 11 deaths (Lavoie *et al.*, 1995). Still, with its focus on smallholder agriculture and urban domestic labour, slavery on St Barth never led to the demographics and economics seen on other Caribbean islands. When slavery was abolished locally in 1846–7, many of the former slaves left the island in pursuit of better economic opportunities throughout the Caribbean and beyond. Today, the population of St Barth traces its ancestry more to the French settlers from Normandy and Breton than to the African slaves whose descendants dominate the population of most other Caribbean islands (Maher, 2013). Even slight variations in ethnic diversity can be the source of much discussion and division in the postcolonial Caribbean.

The social and political differences between St Barth and St Croix did not end with emancipation. Citizens of St Barth today are full French citizens – able to travel, work, and vote as though they lived in Paris or Marseille. While people from St Croix may legally move to the United States where they will be afforded all the rights of American citizenship, those who remain in St Croix stay something of quasi-citizens: unable to vote and not required to pay federal taxes. Passengers arriving in the mainland United States by air from the United States Virgin Islands are required to show their passports. The same demand is not made upon arrival in the Islands.

Today, following these long histories of political, social, and economic inequality, St Barth and St Croix remain very different islands. Owing first to its status as a free port and in recent decades to its niche market for luxury tourism, St Barth has become an island of wealthy individuals with a 2005 per capita gross domestic product approximately 10 per cent higher than that of France as a whole (INSEE, 2005). By contrast, the per capita gross domestic product in the United States Virgin Islands is 70 per cent lower than that of the United States mainland. The economy on St Croix is the lowest within the territory and has accelerated its decline since the closure of the Hovensa refinery in 2012.

Culturally, the two islands differ markedly with regard to the relationships with their respective mainlands; this is exemplified by the issue at hand, the construction of waste-to-energy facilities. Amid a climate of little, if any, cross-cultural controversy, the facility on St Barth was built by Groupe TIRU, a Paris-based firm. The waste-to-energy facility proposed for St Croix was to be built by Denver-based Alpine Energy. Much of the direct action against Alpine Energy and its personnel emphasised their perceived otherness, according to Alpine management. Alpine's proposal was seen as an incursion by an 'outsider' organisation expected to arrive, make a profit through exploitation, and depart (Hurd, personal communication, 24 October 2013) – a scenario not unlike that which played out on the island during the era of plantation slavery and continues to define much of the tourism-based development on St Croix. Postcolonial relationships based on exploitation are especially common on small, resource-poor islands (as evidenced by several cases in this volume). Indeed, such was the contrast in public reception that the current facility on St Barth remains online now, some 13 years after its construction, while the plans for the facility on St Croix appear to have been totally abandoned.

Here, by way of conclusion, I might posit that postcolonial relations with their respective mainlands have contributed significantly to the differential reception of waste-to-energy technology in these two islands. The system of exploitation and economic dependence initiated by plantation slavery in St Croix continues through the dependency maintained by the oil- and tourism-based economy that carried the island into the twenty-first century. When the Hovensa refinery closed, St Croix lost a major sector but still retains tourism. Both modern industries maintain the system by which profits flow from the island to the mainland and the islanders remain dependent upon the mainland for their livelihoods. A waste-to-energy facility developed by a mainland

American corporation would reasonably be seen as merely a further exacerba-
tion of the long-standing inequalities and structures of economic dependence
rooted deep in St Croix's history. Further, the closure of the landfill is being
demanded by the Federal Aviation Administration – an institution of the
mainland United States government, not a St Croix nor even a Virgin Islands,
government body. This example of outsider oversight must have further
encouraged the islanders' resistance to the waste-to-energy facility, regardless of the veracity of Federal Aviation Administration demands. No one can
deny the environmental crisis and aviation hazard presented by the Anguilla
landfill, nor could one reasonably argue that a waste-to-energy facility is not
a more sustainable waste-management system than the landfill. It may be the
case, though, that environmental protection and economic decolonisation in
St Croix represent conflicting goals.

St Barth, on the other hand, with neither the land nor the rainfall required
to establish plantation agriculture, was so economically insignificant to the
French colonial government that the island was traded, not for another colony, but for trading rights in a Swedish port. Upon its return to France, St
Barth remained poor and unproductive, its location and harbour being its
chief assets well into the twentieth century. When tourism did arrive, it came
first in the form of private development, initiated by a handful of wealthy
families drawn to the island for its natural beauty, isolation and potential
for exclusivity, and probably – to be honest – its 'white population [is] very
Catholic and deeply attached to France'. These socio-economic and demographic characteristics have been called by one researcher 'the great originality of St Barthélemy,' who observes of the island's 'persistent quality of
white population' that 'you won't find its equivalent anywhere in the French
Antilles' (Lasserre, translated in Bourdin, 2012, p. 3).

The socioeconomic status and geographic and demographic preferences of
these first St Barth tourists set the stage for what would be a measured development initiative, focusing upon a niche market geared toward the super-rich
(Cousin and Chauvin, 2013). While the economic market for St Barth tourism has since expanded downward somewhat, the restricted growth plan has
proven successful enough to avoid the reliance on mainland developers in
many cases, and when mainland support is required, it has generally been
received on the islanders' terms and under the islanders' direction as the
waste-to-energy project exemplifies. This benign outcome is surely not indica-
tive of French colonialism globally; rather, likely results from a combination
of factors including St Barth's peripheral status during the colonial period
and its economic success as a place for exclusive, luxury tourism.

Epilogue – jewels of the sea

An analysis of energy production, waste disposal, and water supply in St
Barthélemy and St Croix reminds us that the insular Caribbean is a dynamic
region, full of diverse cultures and real-world sustainability crises – not

merely places where carefree holidays are spent. Like the populations of many other practised tourism-based economies, St Barth's residents and leaders have effectively addressed these crises in the background, providing water and waste disposal virtually out of sight to its visitors. The section of the island where the waste incineration powers water generation, ironically named *Public*, is least visited by the public. A high road bypasses the area, connecting Gustavia with the airport and the beaches of St Jean. Quietly industrial, the area of Public privately provides for the truly public areas of the island. Still, from the centre of the incinerator's courtyard, with triaged recyclables stacked to one side and an enormous mountain of combustibles piled in front, it is still possible to look past the waste and glimpse the electric blue water of the Caribbean Sea. Somewhere under that bright surface lies an intake valve, where seawater is pumped onshore to be made fresh, in a process fuelled by the burning of champagne corks and trimmings of bougainvillea.

St Barth is mountainous but small. Clouds created by orographic lifting are often carried west by the trade winds, beyond the island's shore, their precipitation falling uselessly upon the surface of the sea. Even today, when fresh water can be made through desalination, rain falling on the roofs of St Barth is still seen as a godsend – fresh water received without the input of produced energy. I recall once ducking into a tiny jewellery shop called *Bijoux de la Mer*, 'Jewels of the Sea', during a rare cloudburst in Gustavia. The tourists in the shop waited out the shower by trying on expensive Tahitian pearls, but the shopkeepers – members of an old St Barth family – stepped out into the rain, hands raised, whispering '*merci*' for the gift of fresh water. Upon reflection I was reminded of the lyrics to the popular French standard by Jacques Brel, 'Ne Me Quitte Pas': *Moi, je t'offrirai des perles de pluie venues de pays où il ne pleut pas* (I offer you pearls of rain from lands where it does not rain).

The raindrops falling on the roof may have seemed to the islanders more precious than the pearls they were selling inside, but the most valuable commodity from the sea was being produced just over a small hill, through the unglamorous but indispensable process of waste-powered desalination. The same waste, accumulating as it is in the Anguilla landfill on St Croix, contributes to water pollution, not water production. Its attendant flocks of birds and smoke from uncontrolled burning present a risk to aviation and a point source for air pollution. The raw materials are similar. The infrastructural need is the same. The different outcome in waste-to-energy applications on these two islands is not due primarily to any technical barrier but to the discrepancies in culture, economics, and political histories that leave these two neighbouring islands so very far apart.

8   The returning terms of a small island culture

Mimicry, inventiveness, suspension

Jonathan Pugh

Returning terms

Island scholarship increasingly aims to understand islands and islanders on their own terms (Baldacchino, 2006; Grydehøj, 2014; McMahon, 2003; Pugh, 2013a; Stratford *et al.*, 2011). This concern with elevating island voices connects with what many writers from islands have been arguing for years, and includes a call for 'islands not [to be] written about but writing themselves!' (Walcott, 1998, p. 72) and labours by Epeli Hau'ofa (1994) to foreground 'a world of islands' rather than 'islands of the world' (DeLoughrey, 2007; Fletcher, 2011b; Savory, 2011). Indeed, reflecting the salience of the argument, Baldacchino (2004, p. 272), defines island studies itself as 'the interdisciplinary study of islands on their own terms'.

However, the matter of understanding islands 'on their own terms' is far from straightforward, particularly when it comes to islands that have been subjected to colonial rule. For example, George Lamming from Barbados has frequently said that his own terms and words often feel like 'a second-hand bundle of ideas and attitude, imposed and now voluntarily accepted on hire' (quoted in Clarke, 2001, p. 301). In search of new terms and frameworks Barbadian Kamau Brathwaite (1967, pp. 223–4) wrote: 'I must be given words to refashion futures, like a healer's hand.' Similarly, Jamaica Kincaid, Merle Collins, Lorna Goodison, 'Mis Lou' (Louise Bennett-Coverley), Erna Brodber, and Jean 'Binta' Breeze are English-Caribbean authors and poets who have wrestled – and continue to wrestle – with the problematics of inherited terms and language.

The matter of excavating subaltern voices is not, then, merely a methodological or pragmatic challenge to develop better participatory techniques or ethnographies to allow island scholars to hear Caribbean voices, although no doubt these would help. Such a task is allied to a more profound problem associated with feeling the history, pressure, and weight of one's own words. Authors as varied as Walcott (1974), Harris (1999), and Selvon (2006) have written about how island life can be marked by this struggling feeling that the frameworks of reasoning that come out of one's mouth, or that have been marked upon the page, have a certain illusionary and artificial quality to them.

In Walcott's (1974/1998, p. 15) words, what will then go further and deliver the Caribbean island 'from servitude' is the forging of a language that goes 'beyond mimicry', a language that has 'the force of revelation as it invented names for things, one which finally settled on its own mode of inflection'. This is why most great Caribbean literature incessantly turns and returns to everyday Caribbean life, drilling down into it, turning over its terms, seeking ways to return the everyday to us anew through new words.

Whether this inquiry means examining islands as nation-states, complex archipelagos, or other island forms, the concern here is with what it takes to articulate one's sense of self and islandness – whatever that may mean to those engaged in this struggle. In turning to the artificiality of employed terms, such writers point to how the emergence of postcolonial island sub-jectivity is in part this struggle to (re)name the world. Such matters present serious problems for island studies' scholars concerned with studying islands on their own terms. They suggest that in our attempts to understand post-colonial island culture we will need to pay attention to what have become more common themes and tropes such as hybridity (Bhabha, 1994), mimicry (Naipaul, 2012), inventiveness (Harris, 1999), participation (Grydehøj *et al.*, 2015), or resilience (Pugh, 2014) – that often dominate debates *and also heed* less tangible tropes of impasse and suspension (Walcott, 1998). This chapter highlights the importance of some such tropes in the struggle for postcolonial island cultures to articulate life 'on their terms'.

The chapter first focuses upon two publications by St Lucian poet and writer Derek Walcott (1974, 1998) on the basis that these works in particular address tropes of impasse and suspension in Caribbean island life. After explaining Walcott's essays in some detail, I turn to consider empirical research under-taken into one island institution, the Barbados' Landship. Barbados is the easternmost Caribbean country (13.1° N, 59.5° W). A parliamentary democracy, it gained independence from the United Kingdom in 1966; in size is 431 square kilometres (166 square miles); and it has a population of 289,680 (July 2014 estimate; Index Mundi, 2014).

In drawing upon Walcott and the Landship my intention is to work through the significance of mimicry, inventiveness, and suspension. Such concerns are more generally illustrative of my interest in the relationship between Caribbean island expression and life (Pugh, 2001, 2005a, 2005b, 2013a, 2013b; Pugh and Potter, 2001). As well as focusing on Walcott's crucial insights into the Caribbean archipelago, the chapter reports eth-nographic research conducted there over a total of two years encompass-ing 1998/9, 2003, 2011, and 2014, and focused on the Landship and other Barbadian institutions.

The field work involved a range of qualitative and quantitative research processes (Pugh and Momsen, 2006; Pugh and Potter, 2000, 2001, 2003; Potter and Pugh, 2002; Pugh, 2016). With specific regard to the Landship research, I conducted interviews with 20 people, viewed 15 Landship displays, and collected relevant documentation and archival material. I also made detailed observations of how the Landship works in varied formal and informal settings, which included consideration of artistic representations and plays about the Landship in Barbadian society and culture.

Methodologically, the desire to approach both ethnographic fieldwork and research participants with tact is always open to constant refinement, tuning, and balance. This chapter conveys my specific interest in the limits of articulated language and that focus further suggests the importance of being struck by silences. Given Spivak's importance to the discipline of postcolonial studies any mention of 'silence' in a postcolonial context initially sparks recognition of her concern with whether the subaltern can speak. Setting Foucault and Deleuze in her sights, Spivak (1988) rightly criticises Western intellectuals' attempts to rescue the postcolonial 'other' using Western discourses that ironically have the effect of preventing people from hearing the others' voices.

The interest in silence in this chapter, although related, is different from Spivak's own. Instead, as explained below by reference to Walcott and his works and concerns, silence here relates to Walcott's more specific concern for the moments of artificiality associated with the postcolonial condition, when Caribbean contexts reveal how the naming of practices, things, rocks, rivers, forests, and other objects feels borrowed and not owned.

Derek Walcott

Derek Walcott (born 1930) won the Nobel Prize for Literature in 1992. Walcott's (1998, p. 40) concern for what he calls the 'elemental' naming of Caribbean islands is illustrated well in his set of essays, entitled *What the Twilight Says*. This collection, curated by him, will be the main source of inspiration for the balance of the chapter. The set contains Walcott's famous essays 'What the Twilight Says' (1970), 'The Muse of History' (1974), and his Nobel Prize speech 'The Antilles: Fragments of Epic Memory' (1992). The chapter will also draw upon other comments made on Caribbean island culture (particularly, Walcott, 2010) and in the essay 'The Caribbean: Mimicry or Culture' (Walcott, 1974). These works demonstrate how Walcott's career-long aim has been to show how everyday life in the Caribbean is a liminal space that is more than its origins of genocide, slavery, and colonialism (Olaniyan, 1999). For Walcott, Caribbean life is a hybrid space and not purely Black, White, or Indian, but all and more than any colonial language could capture. Walcott thus introduces one overriding tension in his career and critical essays: he rejects the well-worn paths of both Euro- and Afro-centric explanations and seeks uncharted Caribbean territories. Walcott develops two sophisticated themes as a way to elaborate these concerns: 'history as myth' and 'mimicry as the origins of tradition' (Olaniyan, 1999). In this section of the chapter I focus upon 'history as myth' to frame my discussion of 'mimicry as the origins of tradition' in analysing the significance of the Landship.

Walcott (1998) distinguishes 'history as myth from 'history as time', and we can take his explanation in four parts. *First*, in many respects 'history as time' will be a way of thinking familiar to many island scholars and geographers. History as time is linear and sequential. It separates out, arranges and judges people and places according to such categories as 'the modern' (Western) and 'the residual' (Caribbean) 'developed' and 'developing', 'centre' and 'periphery' (Pugh, 2009a). As Massey (2005) writes, when we use terms such as 'advanced' and 'backward', 'developed' and 'developing', we are effectively imagining spatial differences (differences between places, regions, countries) as temporal. Yet, as early as the mid-1970s, Walcott (1998, p. 41) was ridiculing the idea of history as linear time as being 'absurd' in the islands of the Caribbean, instead emphasising the coming together of the Old World and the New World and, in the process, foregrounding tropes of rebirth, renewal, and originality:

> And if the idea of the New and the Old becomes increasingly absurd, what must happen to our sense of time, what else can happen to history itself, but that it, too, is becoming absurd?

He ridicules the idea of 'history as time' because he is instead interested in the immanent 'metamorphosis' of culture, politics, and landscape in the island-chains of the archipelago (Walcott, 1974, p. 5). Walcott argues that it is absurd to reduce the Caribbean to one historical trajectory (Europe) or another (Africa) because in Caribbean culture Africa, England, India, China, Caliban, and Prospero are intertwined into new forms. In this sense, history is not linear 'time' but 'myth' and fiction subject to a fitful muse, to discontinuities and amnesia. Caribbean Carnival, folklore, religion, poetry, creole languages, and patois are all examples of cultural forms that illustrate how the Caribbean refuses the paralysis of history as time. They annihilate 'provincial concepts of imitation and originality' (Walcott, 1998, p. 62). For Walcott then, it a simple matter of 'social necessity' that the 'New World' of the Caribbean rejects the sociological categories of the 'Old' and engages in a more 'elemental' naming (Walcott, 1998, p. 5 and p. 40).

Second, for Walcott the rejection of 'history as time' means something quite particular. In contrast to some Black Power movements, Walcott argues that Caribbean peoples should not resurrect the past, going back in time and crossing over into some 'original' moment; but rather they should focus upon the immanence of the contemporary present where 'maturity is the assimilation of the features of every ancestor' (Walcott, 1998, p. 36). Walcott is often contrasted in the Caribbean with the Black Power movement of the 1960s and 1970s associated with Coombs, Pantin, and Rodney (Olaniyan, 1999). Walcott (1998, p. 37) argues, however, that such radicals' 'servitude to the muse of history has produced a literature of recrimination and despair, a literature of revenge written by the descendants of slaves or a literature of remorse written by the descendants of masters'. Many radicals in the Caribbean continue to reduce Caribbean language to a form of 'enslavement' so that their 'admirable wish to honour the degraded ancestor limits their language to phonetic pain, the groan of suffering, the curse of revenge' (Walcott, 1998, pp. 37, 39). Provocatively, he argues that 'history is irrelevant' to the contemporary Caribbean – not because it has never mattered but because what matters in a region at the crossroads of multiple spatial trajectories is 'the loss of history, the amnesia of the races' (Walcott, 1974, p. 6). In what has become a famous passage from 'The Muse of History', Walcott (1998, p. 64) sheds light upon his distinctive perspective:

> I give the strange and bitter and yet ennobling thanks for the monumental groaning and soldering of two great worlds, like the halves of a fruit seamed by its own bitter juice, that exiled from your own Edens you have placed me in the wonder of another ...

Third, for Walcott this gift means that Caribbean islands are diffracting spaces where it is often impossible to distinctly separate out multiple trajectories and inheritances, and again, this means something quite particular in his work. Walcott shows concern for how attendant themes of creolisation and hybridity have too easily been reappropriated by European and North American scholars in recent decades, becoming shorthand for creative cultural adaptation and even postmodernity (see also Palmié, 2013). By contrast, for Walcott (1998, pp. 62–3) the struggle to express the distinctive originality of Caribbean landscape and culture requires a 'thrilling metamorphosis', where language itself becomes an integral 'living element' in the process of articulating the archipelago on one's own terms. Yet, as Walcott (1998, p. 42) says:

> Most writers of the archipelago contemplate only the shipwreck, the New World offers not elation but cynicism, a despair at the vices of the Old which they feel must be repeated. Their malaise is an oceanic nostalgia for the older culture and a melancholy at the new ...

Fourth, step-by-step movements through themes of 'history as time', mimicry, and invention then reveal the central thrust of Walcott's work as it concerns me in this chapter. Far from denying history, as it might at first seem, instead Walcott incessantly rakes over, recalls, and recounts his inherited terms, and finds them wanting. Documenting his early years as a playwright, he writes that such feelings of impasse could become overwhelming, so that 'the sense of hallucination increased with the actuality of every detail' of the island that he was trying to survey (Walcott, 1998, p. 33). Here, for Walcott, it is important to give serious and considered attention to such feelings; to dwell in them at times, and to feel the oppressive weight of one's own inherited terms. He reflects upon such moments of contemplation by entitling his most famous set of essays *What the Twilight Says*, in the opening pages of which he refers to the twilight as 'a metaphor for the withdrawal of empire and the beginning of our doubt' (Walcott, 1998, p. 4).

Later, he describes the twilight as a suspended experience (Walcott, 2010). So when Walcott (1974, p. 4) characterises the Caribbean as the 'spiritual force of a culture shaping itself' he is engaging the inventiveness of creative adaptation, and constituting Caribbean islands as diffracting spaces, composed as multiple and relational trajectories. Important as these are, however, Walcott (1998, p. 40) is also calling Caribbean people to arms by emphasising the struggle for the 'elemental privilege of naming the New World', and focusing upon themes of 'self-annihilation, to beginning again'. There is newness to the Caribbean associated with the 'loss of history' and 'the amnesia of the races' (Walcott, 1974, p. 6) – something that great writers of the New World more generally have had to contend with – Neruda, Whitman, and the early Perse and Césaire among them. As one of Walcott's greatest admirers, Wilson Harris (1998, p. 31) also notes, to 'arrive in tradition involves an appreciation of profound tension between originality and tradition'. Here Walcott in particular encourages a slowing down, a recounting and raking over terms used to constitute everyday life. In doing so, Walcott demonstrates how creative and critical writing are often mutually informing in Caribbean literature, because they refuse the separation of analysis and invention. He profoundly illustrates the feeling that new words need to be minted and old words and frameworks of understanding stretched and revised (Lazarus, 2011, p. 83). The Caribbean institution of the Barbados Landship exemplifies the importance of such concerns.

The Landship: mimicry, invention, suspension

> The Caribbean sensibility is not marinated in the past. It is not exhausted. It is new. But it is its complexity, not its historically explained simplicities, which is new.
>
> (Walcott, 1998, p. 54)

Unique to the island of Barbados, Landships are voluntary neighbourhood associations and frequently a source of pride for members of Barbadian society (Allsopp, 1972; Burrowes, 2005; Carrington *et al.*, 2003; Downes, 2000; Fergusson-Jacobs, 2013; Meredith, 2003, 2004). Rather than a static thing, the Landship is better perceived as a moving performance (see for example, a 2008 YouTube film entitled 'The Barbados Landship').[1]

Historically, Landships have provided welfare and social and organisational structure for the poor, particularly during the depression years in the 1920s and 1930s. Today, as an institution the Landship is an amalgam of different parish Landships and, although now mostly in demand for their cultural performances (elaborated upon below), in the past Landships supplemented the vitally important social security functions of Barbados' Friendly Societies. Those Friendly Societies represented a 'network of grass roots organisations' that provided workers and their families the opportunity to articulate collective concerns; they were the 'bedrock of political organisation and development' during the years of Depression (Hunte, 2001, p. 134), with enforced savings being made to provide insurance and social security for members. Supplementing the Friendly Societies, the Landship was a 'powerful social factor in the lives of its members [and] represented a commitment to self-help and survival within the working classes' (Griffin, 1997, p. 97). As well as functioning as a form of entertainment for Barbadians, as I explain shortly, the Landship effectively served as a welfare organisation and credit union for many poor Barbadians in the run-up to independence from Britain in 1966, a pivotal time in Barbadian history. In the 1920s, there were around two Landships in each of Barbados's 11 parishes. St Michael, where Bridgetown – the capital – is based, had many more and probably included six Landships. Indeed, by the 1930s 'there were three fleets of sixty ships with membership of over 3,000 men and 800 women' (Fergusson-Jacobs, 2013, p. 4). Signifying the Landships' significance, when Pan-Africanism leader Marcus Garvey came to Barbados in 1937 it was the Landship that was singled out to act as Garvey's guard of honour for the short period of his stay on the island. The key point here is that the Landship is an important expression of Barbadian culture and means of rearticulating public space (Burrowes, 2005; Fergusson-Jacobs, 2013; Meredith, 2015). There follows three points about the details of the Landship as these specifically reflect key themes in the work of Walcott noted above, concerning mimicry/mockery, inventiveness, and suspension and impasse.

As Walcott (1974) explains, as part of the process of suppressing Caribbean religion, carnival and many folk traditions are frequently reduced to surface spectacles and sidelined as mere forms of mimicry or, more 'radically', mimicry/mockery. It is as if they are given a certain begrudging credit for orchestrating a limited amount of mocking freedom via the slippage of European and African rules. As Brathwaite (1971) and Henke (1997) remind us, within Caribbean societies themselves festive moments of self-realisation are too often relegated to such tantalisingly marginal spaces. But whenever they become too overwhelming, the governing ethos lashes out or tries to 'co-opt them by turning them into empty rituals adorning their self-congratulatory society events' (Henke, 1997, p. 55; see also Hosein, 2012). As I now explain, such themes have also run through the history of the Landship.

It is widely accepted that the Landship was started in 1863 by Moses Wood, a sailor serving in the British Navy. But its history is far from clear. Wood may have been called Moses 'Ward', and may have been a white Englishman, a coloured Barbadian from Cardiff and Southampton, or a sailor who was a returnee Barbadian national. However, Fergusson-Jacobs (2013, p. 4) has argued that it is 'not feasible that Moses Wood was the originator, since his records show his birth as 21st of January, 1860'. Fergusson-Jacobs (2013) instead notes that the Landship probably existed long before the 1860s and likely was practised in some form within plantation communities by African slaves. Although the reason is lost in history, the Landship was formed by re-enacting, exaggerating, and developing ship's activities on land (Best, 2001). Subsequently, as the movement grew each 'ship' that formed in the parishes was named after a British vessel such as the Iron Duke, Victory Naval, and Nelson (Burrowes, 2005). As in the Friendly Societies, 'members were ranked and defined according to the status hierarchy used by the British Navy' (Griffin, 1997, p. 97). Today they often wear naval uniform and women frequently dress as 'nurses' or in domestic servant uniforms akin to those worn approximately 50 years ago. Meetings ('susu') take many forms, are often musical, and frequently involve prolonged public addresses by different members. Until World War I, only men participated.

Griffin (1997, p. 97) argues that the Landship 'fulfilled the need for order and respectability, and mutual assistance among the unemployed poor during the depression'. However, for others the Landship can be seen as mimicking/mocking traditional colonial dress and music (Burrowes, 2005). In addition to the aspects of the Landship noted above, and which replicated colonial and ship activities on land, the organisation often involves Bajans (Barbadians) using polysyllabic words, accentuated gestures, and actions. The vice-chairman of a Landship can introduce the chairman as, for example, being 'fundamental and groundamental enough to take and hold his place like Mark Antony or Cicero, or Blake or Drake, or Wordsworth' (Collymore, 1955/1992, p. 107). But Landships cannot simply be reduced to mimics of British culture because they are also accompanied by 'Tuk' music, believed to be taken to Barbados by the slaves from Africa in the early 1600s. The name 'Tuk' is probably derived from the Scottish word 'Touk', which means to sound or beat a drum. Instruments include tin flute, kettle drum, and bass drum. Tuk music was prohibited by British colonists and did not emerge overtly until after emancipation in 1838. Its introduction into the Landship was not then a simple reintroduction of a lost cultural practice, but illustrates how oppression and invention are intertwined in Caribbean culture (elaborated below). Tuk music was reinvented by means of its prohibition and then probably renamed with a Scottish term. Similarly, certain naval practices were adapted from colonial masters and also took on new forms. What resulted was a new and distinctively expressed Barbadian cultural form.

My first point about the Landship therefore is that initially it can appear to demonstrate Bhabha's (1994) point that if colonialism takes power in the name of history it repeatedly exercises authority through figures of farce and traditions rich in irony, imitation, and repetition. For Bhabha, the menace of mimicry/mockery does not result in a new social order but comes from the partial representation and recognition of the colonial object – almost, but not white. Hence, the Landship mimics/mocks colonial dress: it does not foreground an exercise of dependent colonial relations through narcissistic identifications, so that only the white person can represent the black person's esteem, and the black person stops being a full 'person' as a result. Rather, there is a constant slippage of rules (Burrowes, 2005). Hence, too, members of the 'ship's crew' of the Landship regularly appear to execute incorrect ship manoeuvres during performances (for example, the marching step of the right hand, right foot). This mockery of colonial practices is accentuated and dramatized by the Landship 'Instructor', who gives the appearance of becoming highly frustrated by the inability of the ship's crew to 'get it right'.

But Bhabha permits only limited consideration of the trope of inventiveness, a constraint that takes us further and closer to Walcott's concerns outlined earlier. Olaniyan (1999, p. 205) argues that there is a 'profound difference' between Bhabha and Walcott. The former focuses upon the slippery nature of inherited colonial discourses so that the resemblance and menace of mimicry in his work can only result in endless compromises and, unlike that offered by Walcott, there is never a new horizon or order. Even when Bhabha (1994) introduces the notion of 'hybridity' there is nothing essentially anti-hierarchical or subversive about it. By contrast, Walcott (1974) is more concerned with transcending old orders. Walcott's 'mimics are conscious agents with a dire if incoherent sense of need, Adams impelled to conscious invention' (Olaniyan, 1999, p. 205).

The Landship performance illustrates such concerns well, for historical periods, characters, and epochs are expressed without regard for sequence. The idea of 'history as time' in the Landship becomes both warped through mimicry/mockery and absurd (in Walcott's sense, as noted earlier). The Tuk rhythm accentuates the transformative vibrancy and spontaneity still further, so that once the Landship reaches 'full stream', 'rhythmic improvisation and expressiveness of dance take over' (Best, 2001, p. 53). As one participant in my research described it, the 'Landship expresses a Barbadian essence [that] transcends old terms' and the frameworks of meaning that could be used to describe it.

To say, therefore, that the Landship is merely a form of mimicry/mockery now feels a rather reductive form of analysis; this then, is not about reducing tradition to mimicry/mockery – seeing these as the parameters and boundaries of culture. But rather, as Walcott (1974) suggests, it is about mimicry being seen as but the starting point for questions of rebirth and renewal. And as Walcott (1998, p. 40) later expands in a particularly Nietzschean moment, the expression of Caribbean island culture is about a 'belief in a second Adam, the re-creation of the entire order, from religion to the simplest domestic rituals'. Thus, oppression and invention are particularly closely intertwined in the Caribbean. In the Landship, inventiveness is directly related to the banning of slave drums and their emergence in new Tuk forms. Inventiveness emerges from oppression which is intrinsic to the Landship as a form of resistance and new culture. Lovelace (2013, p. 70) similarly describes the stakes of invention and political resistance in Caribbean islands as follows:

> The struggle for the survival, expression and innovation of the cultural forms that emerged out of people's struggle to declare themselves against a dehumanising system, and for a different vision of what it is to be human, became, in effect, also the political and economic struggle.

As for inventiveness, along with so many other cultural forms (Lovelace, 2013), the Landship is often reductively characterised as 'the Other' in Barbadian society. Burrowes (2005) effectively traces how this othering was achieved from the earliest years of the movement and its practices. In 1931, *The Advocate News* reductively called the Landship 'Little Englands [*sic*] Navy' (Burrowes, 2005, p. 216). Newspaper reporters wrote that Barbadians should not 'blush' at the sight of the Landship because it was installing British values of discipline (ibid., p. 216). Certain writers, no doubt reflecting the view of many but not all Barbadians today, have reduced the Landship to a 'masquerade' (Marshall, quoted in Burrowes, 2005, p. 228). Thus, today it is difficult to discover the Landship on its own terms; this is because, as one Landship participant told me, it has 'already been discovered as something else'. The Landship has been reduced to a state of mimicry that at best plays with the slippages of inherited cultures from Africa and Britain, rather than creating an inventive expression of politics in its own right. This is not to say that the Landship would now necessarily be a radical political force for change. Rather, the Landship historically demonstrated 'the organizational capacity of the working class ... [and] helped it develop its political consciousness, thereby demonstrating that workers were prepared for the challenges of organized mass political activity' (Griffin, 1997, p. 97).

Today, as with many other forms of culture in the Caribbean, including religion, carnival, and folk resistances, the potential to push the invention further is effectively suppressed and silenced – there is both impasse and suspension. On 26 April 2014, while in Barbados, I watched a play entitled 'The House of Landship'. Written and directed by Winston Farrell, this play explores several complex and interrelated themes concerning what it means to articulate an island 'on its own terms'. The play is about a family involved in a Landship called the 'Bim' (a colloquialism for 'Barbados'). The central figure of the family is female; a mother who stands effectively as the head of the Landship and for the ideals of island independence. One of the key plotlines of the play concerns whether or not the government will scrap the Landship parade from Independence Day celebrations and highlights the mother's centrality in arguing for its inclusion. A related plotline focuses upon how different members of the family – a son and uncle and their particular associations with the Landship – come to symbolise relations between those who stay on the island and those who leave and then return to the island after travelling abroad. As an illustration of these dynamics, the uncle leaves Barbados only to return years later to support an oppressive building project that would effectively see the destruction of the 'dock' and home of the Landship. Throughout the play, the son also wavers about whether to leave Barbados and the Landship for a 'better life' abroad. There is, however, finally a 'happy ending' to the play – the son decides to stay and support the Landship, the uncle changes his mind about the destruction of the 'dock', and both re-join the Landship family. At the end of the play, the House of Landship triumphantly stands for a uniquely independent Barbados, and this is expressed in how the Landship is finally included in 'Independence Day' celebrations. The play demonstrates how the Landship continues to hold a firm place in the popular Barbadian imagination. It shows the importance of renaming and re-enacting the island imagination, and emphasises how the Landship continues to throw up important questions concerning what it means to articulate the island on its own terms.

Suspending the conclusion

As the editor of this book has pointed out elsewhere, in contemporary island studies islands are increasingly constituted as relational spaces, imbued with meaning from multiple perspectives and trajectories (Stratford, 2003, 2008; Stratford and Langridge, 2012; see also Tsai and Clark, 2003; Clark and Tsai, 2012; Baldacchino and Royle, 2010; Pugh, 2016). But in this chapter I have sought not so much to foreground plurality as to consider tropes of mimicry, inventiveness, and suspension in the struggle by islanders to speak of and for themselves 'on their own terms'.

While I have focused in part upon Walcott, my sense is that such concerns are both more and less applicable to other contexts. On the one hand, Walcott (1974) himself continues a much longer tradition of struggling to discover the 'Americas' on their own terms. My underlying sense here is that Walcott continues a particular tradition developed in the early 1800s by Ralph Waldo Emerson (2015) in New England, and specifically the quest to understand how 'part of the task of discovering philosophy in America is discovering terms in which it is given to us to inherit the philosophy of Europe' (Cavell, 1996, p. 348). Like Emerson, Walcott has regularly been accused of adopting the conventions of European writers. But also like Emerson, his argument is that he was irretrievably given his language, frameworks, and terms, and so can no more give them back than others can now claim them for their own.

The question for both Emerson and Walcott, therefore, is how can inherited terms be used to grasp to the everyday as they experience it (which may mean inherited terms not grasping that experience effectively at all). As Loreto (2009, p. 12) notes in the only other publication I know that makes the connection between Emerson and Walcott, both authors stand on 'the border of what he has not yet been able to express because it is constantly changing into a new shape'. If there is one characteristic that aligns Walcott and Emerson it is this rejection of 'History' as a reified or essentialised category, in favour of an incessant turning, to name things anew with inherited terms. These great writers encourage readers to take seriously the possibility that a state of suspension has been reached and which has come to characterise the everyday. Indeed, throughout the key critical essays contained in *What the Twilight Says*, Walcott dwells in these feelings of suspension that come about from one's sense of self and nation being reduced to the product of elsewhere. Such themes of restless dislocation are prominent in postcolonial literature, particularly that of the Caribbean: as Noxolo and Preziuso (2012) point out, they are illustrated by tropes such as the ghost, the mad woman, the symbolic city, or the ship, but they have been given less attention in recent years.

In passing, my sense is that the rise of philosophical traditions including pragmatism across the Western social sciences and humanities in particular, and the pragmatic drive to 'do something quickly' about such urgent (island) concerns as climate change and natural disasters, has moved less tangible concerns aside for the time being. The irony then is that while 'participatory' initiatives on islands are being conducted by international development agencies at an incredible rate, they continue to be initiated on other people's terms (Pugh, 2013b; also Dean *et al.*, this volume).

As a final point, Walcott (1998) refers to Chamoiseau's novel *Texaco*, a work about the ironies of Caribbean planning institutions. Walcott describes the novel as a sound exemplar that runs up against moments of impasse and suspension in island life. In Chamoiseau's writings a carefully placed 'hear dis', 'you listenin?', or 'I could finish talking', or, in French Creole, 'moin' or 'tends ca', subtly illustrate a 'delightful scepticism' – they demonstrate how creole is an expansive syntax often dominated by rhetorical questions and heightened gestures that express the 'continuous bemusement at the ironies and stupidity of fate, of history of orders laid down' (ibid., p. 226). For Walcott, such phrases do not pepper the novel to give it local flavour; instead, these expressions are 'organically, the texture, the heat of the meal itself, the crunch of the yellow yam, dasheen (taro), white yam, saltfish, and maki (bark beer)' (ibid., p. 227). Such phrases are carefully crafted and purposeful in acknowledging the agonies of an entire race and archipelago, and the struggle to articulate life on and with their own terms. They are not mere asides but rather encourage the reader to inhabit the impasses and interstices themselves; provoking us to slow down, suspend, and pay attention to the magnification of ordinary events, gestures and movements. Chamoiseau's gestures – 'hear dis" and 'you listenin?' – do not mark time as if it is a movement forward. Instead, they make time, holding open the present to attention and the possibilities of unpredicted exchange. They invite us to consider how the study of islands 'on their own terms' can also be about slowing down, exploring the impasses and feelings of suspension, and contemplating the silences.

Note

1 www.youtube.com/watch?v=JLUXkUK_GFU.

9 Conversations on human geography and island studies

Elaine Stratford and authors

Preamble

Like many edited collections, this volume had its genesis in conference presentations and ensuing conversations, and that has been explained in the first chapter. But what edited collections capture less often is that these conversations may extend beyond the boundaries of a given conference venue and may involve more than the relatively discrete communications between one or more editors and several authors who produce individual chapters. I have wanted to unsettle the 'how' of the edited collection in two ways, and this and the next chapter are evidence of that aspiration – but I have explained much of that motivation in Chapter 1. Here, let me just remind the reader that this collection is intended to be read as a group effort, an ongoing exploration among colleagues.

What follows, then, is the edited transcript of a synchronous, recorded telephone conversation in October 2015 that was held over several time-zones spanning from Hawaii to Tasmania via the United Kingdom. The two-hour conversation was prepared for in two ways: first with all of us reading each other's chapters (with the exception of the introduction and Royle's retrospective and prospective views which follows this chapter); and second with each of us then generating questions for a collective discussion on the basis of that reading. Those questions were then circulated in advance of our teleconference call, and many of them were covered in the discussion that unfolds below.

The ensuing conversation provides evidence of our intellectual and practical struggles; of the insights we derive from our individual work and from working with each other; and of the questions that these labours have either affirmed, unsettled, or indeed generated. This conversation has also enabled us to consider our positions as scholars who straddle interrelated disciplines and methodologies, chief – but not solely – among them human geography and island studies. And our time together has confirmed that new or newly configured forms of collaboration exist when we make use of communication technologies to deepen the ways in which we co-produce written works arising from our conversations at other, earlier times. By such means, each of us has been energised and engaged anew.

ELAINE: I want to thank you all for giving time to this labour, and will start by asking what attracted each of you as a geographer, sociologist, or other, to studying islands?

DAVID: My background is in geography and history. In my postgraduate studies, I did a dissertation on the imaginings and perceptions of the sub-Antarctic Auckland Islands and, as I was doing my literature review, I found a copy of Stephen Royle's *A Geography of Islands*. I did some further searching and found some island studies journals. I became fascinated by islands and the study of islands, and this grew with my Master's and then my PhD, which was on the management of cultural heritage and natural protected areas on select uninhabited offshore islands of New Zealand. What really interested me was the mysteriousness of those islands, and the question of how islands are seen as binaries – prison and paradise, for example.

RUSSELL: I am also a geographer by training and, much like David, I was attracted to islands rather early in my studies. During my Master's, while I did not seek an island case, the work I was doing was based on three small islands very close to my home town in Tampa, Florida. Much of the literature I was referring to was coming out of Prince Edward Island from people such as Godfrey Baldacchino. He and I developed a back-and-forth relationship, and I was then invited to apply to spend a year at the Institute of Island Studies at the University of Prince Edward Island where he worked and, by the time I was finished with that year, I had been converted to a full-time island scholar.

THÉRÈSE: I think that my involvement in island studies was a little accidental. I moved to Tasmania, an island, to study. I did a Bachelor of Arts with a major in geography, and island studies was an important part of that. But, I also did a philosophy major and became particularly interested in Heidegger (primarily because I violently disagreed with him on a number of levels and wanted to understand why). Thus, I was also interested in phenomenology and philosophy of place and this interest is reflected in my chapter [in this volume]. It struck me that place, theories of place, and islands studies theories seemed to overlap in the sense that often almost exactly the same words were used to describe both. I wanted to see what theory and philosophical ideas looked like 'on the ground' together. More broadly, I am interested in the ways in which we use islands and use the idea of islands, with particular reference to the Australian historical context – islands as prisons and, now, as carceral places for asylum seekers. So, for me, there is a social justice aspect to these inquiries. I am trying to understand what it is we are doing with islands, and to reflect on how we are mobilising the idea of islands to do it. So there is both an element of intellectual curiosity and a practical motivation in my attraction to studying islands.

ELAINE: Marina, what is your particular take on these sorts of issues as a sociologist with a particular interest in social and spatial justice?

MARINA: My dissertation and all of my research have been on islands. I remember in the mid-1990s telling people that we needed to study islands. I wanted to study islands but there was nothing available in sociology. There still isn't much. I then Googled 'island studies' sometime in the last three to four years, and found an entire world of islands and island studies. In keeping with what Thérèse said, thinking about social justice and social movements using the island as framework is really interesting, and an important and practical application of what we contribute as island studies scholars. Grant McCall says that islands are harbingers; we think about islands, we think about what we have been thinking about on islands, and we think from the metropole as island studies scholars and hope to influence continental thinking as well; there is a leadership role here for us.

ELAINE: Using the term nissology, McCall argued that we should study islands on their own terms. So can we open up the conversation – not only to what attracted you to the field of island studies but to the question of how do you study islands on their own terms?

KATE: I am still new to island studies and came upon it as a geographer dealing with ocean studies. My projects as a geographer have been to disrupt the privileging of static, stable, land-based Cartesian thinking, and for me islands are practical in terms of studying oceans because so many parts of the ocean are managed by island governments and island peoples. But, in addition, thinking through islands is really useful for helping to up-end the binaries of land and water, and to reorient geographic thinking about social ocean spaces. So for me, island studies has been really useful for ocean studies – the two go hand-in-hand.

JON: What attracts me to island studies are questions of justice and I think that is what brought me to this area of inquiry: questions of justice and of what it means for people from islands to articulate their identities on their own terms? I came to island studies initially because I was very interested in the ways in which 'development' works, and in the ways in which academics and consultants engage with development on islands. I was interested in how difficult it is to engage in development processes that are participatory; to allow people to articulate who they are on their own terms – precisely because so much development is not framed in that way and is driven by outsiders; so, too, are development programs and consultancies. Thus my interest has been oriented around practical questions about that tension and not least when considering questions of social justice. Yet, it feels problematic and difficult to engage in such questions as an academic based in the UK – to be working somewhere such as the Caribbean.

ANNIKA: That sounds very similar to my experience. I was attracted to the study of islands almost incidentally. I studied development studies in human geography. I have long been interested in, and concerned about, the impacts of climate change on people and biodiversity around the world – coming from one of the world's largest coal ports [Newcastle, Australia]. The first time I studied islands was in the Torres Strait, where I was looking at responses to climate change at different government scales and using a postcolonial theoretical lens to look at that in action on climate change responses from the federal and state government. Later I chose to study in Kiribati because I wanted to look at how climate finance was affecting the ability to adapt to climate change in a specific place. In broad terms, that place didn't really have to be an island, but I chose an island: I chose Kiribati because it is archetypal as a 'climate change place' and because it was potentially poised to receive a lot of climate finance, being a small island developing state and least-developed country. I think the fact that Kiribati is bounded and small subconsciously appealed to me as a way to scope down my research and understand these dynamics in a place. I feel as if I also came to understanding islands on their own terms inductively: it was about observing how things were happening; how development practitioners were behaving; and how development projects – or one particular adaptation development project – was unfolding. Continental discourses are often taken-for-granted and they render islanders constantly as deficient, and islands as constantly (re)created spaces for intervention. Perhaps in retrospect, there was concern for what I might now see as social justice – trying to see Kiribati for what it is without any reference to a continent or continental thinking. And all that helped me to be self-reflective in my own research practice.

ELAINE: Russell King once asked the question who is an islander? The answers remain fraught, not least because of different forms of migration. So, there do seem to be ongoing tensions informing the debate about whether or not it is possible to study islands on their own terms, and – indeed – for what ends we engage in such work. So could we turn briefly to explore what you have seen as the desired or expected outcomes from the studies that you have done?

DAVID: As a matter of fact, I don't usually use the phrase to 'study islands on their own terms' and I sometimes struggle with what it means: perhaps, most succinctly, it means empowering islands and islanders and correcting the misinterpretations people have of islands, and that certainly is my desired result or outcome and motivation for this work: knowing that islands are not as simple as they may seem and so not as vulnerable or as isolated or as bounded as people may think of them. In terms of my research, this means seeing and making it clear that islands are not completely natural or pure but that they also have complex human histories as well.

MARINA: A key motivation for me relates to islanders and mobility, and I think of Hau'ofa's work and the idea that islanders are mobile; and that these practices of moving are part of the character of being an islander. That raises the question of whether it matters where you are located? For example, I am teaching a new island studies course right now and am so thankful to be teaching it on an island with a student body whose members have long histories as native Hawaiians and people who have migrated here for various reasons. Or consider news from Lesvos in Greece, and suggestions that cruise ships had ceased berthing there because of the migrant crisis in Europe, and think about what that will mean for the community there, even for a short time. Location matters, and these particular dynamics brings to public attention the need to think about islanders and island conditions.

ELAINE: Such matters are indeed key concerns for human geographers and I would be interested for us to tease out some of the ways in which your work makes substantive contributions to the discipline. For example, Jon, when you go to the Caribbean what are you hoping to achieve from your studies and methods of study in those island places?

JON: I think that people who work on islands and in island studies feel the weight of this responsibility quite heavily; feel the need to question the ways in which development works and is done, and consider how it frames problems and solutions. But I wanted to add two points to the question of movement and the study of islands on their own terms with regards to stakes that are involved, at least in terms of my own research. First, there is a tension in island studies – which I think should remain in play, but which I think is often over-accentuated – that islands are either isolated or that it is all about movement. I think it is actually both. Islands are the products of movement, a coming-together in a particular locality. So, it is not that there is a relationality on one hand and then an isolation or insularity on the other. As people such as Glissant or Walcott and other island writers relate, insularity is actually a product of different comings-together and movements that are precisely unique to island contexts because of all the different movements, flows, and relationalities that are put into them. I think that was also perhaps what Marina was hinting at. The second thing about the study of islands on their own terms in relation to what is at stake in my own work in the Caribbean in colonial contexts is this: it is also about what it means to discover one's own terms. When you are emerging from 400 years of history and slavery in the Caribbean what would it even mean to find your own terms?

ELAINE: Russell, how does Jon's observation sit with your own experience of working in the Caribbean with different islands because, of course, the Caribbean is not one place, and rather more like a multiverse.

RUSSELL: That is true. There simply is not one set of terms per island especially in the Caribbean or the Pacific – places that have been layered with various waves of migration and then, of course, with the effects of colonialism. Each of those cultures or peoples have brought with them their own terms and when they have layered upon and integrated within one another neither realities nor representations are as straightforward as we island studies or island studiers may wish them to be. We say we are going to study islands on their own terms as if there were a cohesive body of terms to consider in the first place; I am not sure there is.

JON: I think that this is where the archipelagic approach has become interesting in island studies lately, because I think it often tries to bring out that precise point.

RUSSELL: Before we get too far from Marina's comments, I wanted to say that I really appreciate hearing her use the exact same term for people who are arriving in the Greek islands as that which we hear on the news every day and among non-native students whom she teaches in Hawaii. To hear them both referred to as migrants really levels the playing field in a way that I'm afraid much of the media and much of our conversation fails to do.

ANNIKA: Thinking about studying islands on their own terms, and about whether that means that only islanders can articulate those terms, I feel that it means more than that. It does mean trying to provide a platform to hear islander voices and narratives but it is also about trying to see the place as an environment on its own terms. I see Russell's point and I don't know if it is possible, but I think that it is worthwhile for islanders to try and see islands on their own terms. I think outsiders still tend to dominate island studies, but I think that they can make valuable contributions. And I recall an article about this question by Baldacchino that was written about 2008. He was suggesting that some of the social effects of islandness are that it can create both conservative attitudes and egalitarian approaches – with people not wanting to show off or stick out, nor wanting to pass critical commentary. I think you see that in Kiribati. It means that there are a lot of barriers to people being able to articulate their experiences, and a lot of islanders seem to go other places in order to be able to write about their country, their island. So, I think that there is a place for outsiders. I think that they can try to articulate ideas about islands on their own terms. I think that this practice develops more perspectives and richer perspectives. It is not just about islanders speaking for themselves but about everyone trying to understand islands as they are and celebrate them as they are.

THÉRÈSE: It strikes me as, that just as isolation and mobility actually coexist – for they are not one thing or another – so there is a multiplicity of islands terms. It is not so much about Island Terms: it is more about the terms of each island. I think that is part of my motivation in trying to understand islands through place and particularity: the terms of and for Bruny Island are not going to be same as terms of any other island.

ELAINE: And the terms within Bruny Island are not going to be the same either? The divisions within islands certainly are well documented.

THÉRÈSE: Precisely. There are issues of scale here too. We talk about 'islands' but we are talking about very, very small places and about very large places – and there are huge divisions in what those terms mean. As you know, Elaine, there are people who live in the middle of Tasmania – an island – who have never seen the water. I am not always quite sure what understanding islands 'on their own terms' means either: but one of the reasons that I am interested in doing so is to understand how we impose ideas of islandness onto islands. Sometimes we don't even know that we are doing that; it is subconscious. For us, to be able to identify our own terms – the terms that we are using and consciously and unconsciously imposing – we need to be able see the other side of that: in order to separate the two one needs to be able to see what is in place. I don't think that imposed or inherent ideas and terms are discrete either. I don't think that they are any more discrete than mobility and isolation. I think that they are intertwined and that they co-constitute each other. People who live islands are not immune to the idea of 'island' in framing their experience – and some of those ideas in fact rise out of islands themselves. So I am interested in how such ideas and terms are imposed on islands, arise out of islands, are intertwined, and play out and, getting back to questions of social justice, I want to know what effect that has on people and places.

ELAINE: And thinking about notions of mobility invites us to ask how ideas and individuals are mobilised and emplaced in different situations. Now, can I ask how each of you would describe your approach to the study of islands, in this volume or in other studies? And how would you describe the ways in which your epistemologies and approaches to islands have emerged from the study of islands or whether applied to islands from without? Is it possible to study islands on their own terms through the lens of a continental or other epistemological framework?

MARINA: Well, I have gone to many sociology conferences and the absence of islands has been notable. So, I started thinking 'let me propose the sociological significance of islands and ask why are we not thinking about this?' Recall that I work with feminist frames and use an intersectional approach considering several social inequalities. I also see islandness as an important intersection: something that, at least in the sociological literature, hasn't really been considered. What does it mean to be from an island or to live on an island? My approach deals with both people who are essentially brought up on islands and with people who arrive.

KATE: I am still fairly new to island studies, so I am very open to learning from the discipline and from the experience of doing research on island sites. I envisage an approach to studying islands on their own terms in my own research as being an iterative approach in which I have to understand my own positionality when conducting interviews or trying to gain access to archives. I feel the need to stay open to possibility.

ELAINE: Do you think your approach would be different if you were doing your study in the middle of Arizona?

KATE: That is something worth trying to understand as well: thinking of islands as a category without trying essentialize islands. I am not from

Arizona, though I work there at present, and drawing from human geography and considering epistemology and positioning, I think that there are similar strategies to any kind of investigation when you are coming from outside. Perhaps it would be different – but I don't know.

JON: Interesting. Most of the papers that I have written have started off from a situation. For example, I have been looking at how an institution develops in the Caribbean, or how a process of participation works, or some other situation like that. I have always started off from a kind of moment of impasse in the lives of the people who were involved in that process. So in a specific way I am interested in this question of islands on their own terms, in a moment in a participatory development process, or in a moment of the development of a sustainable development ministry or something like that, where the terms themselves have been thrown into question by the people in that institution or in that process.

ELAINE: Again the same question applies: acknowledging that the British Isles where you live are, indeed, isles, does that situational approach emerge from working on islands or was that something that you took with you from the other work you had been doing on democracy, for example?

JON: I think both. I think that in life we all hit a moment where we are like 'My gosh, are these my terms? Am I speaking for myself?' I think Heidegger was mentioned earlier. I disagree with Heidegger on lots of things too, but the broader question of what does it mean to speak for one's self is such a 'human question', and in island studies, and particularly in island literature in the Caribbean, it is a recurring tension.

ELAINE: David and Russell. Do you have anything you want to add?

DAVID: I think the idea of place greatly influenced me as part of my approach to both human geography and place; in particular how location influences place – so think of the influence of islandness. I think that this is the centre of quite a lot my research: focusing on the characteristics of islands – isolation, separateness, boundedness – and thinking about how islandness affects what happens on islands and how islands are perceived.

RUSSELL: I can see how that works, and scale is another concept from human geography that I would throw in to the mix. Especially for me with my work on infrastructure, questions of scale are at the forefront. The small scales being dealt with on islands are often not usual for mainland organisations, whether they are NGOs, development organisations, or companies and that often serves as one of the sources of discontinuity on these projects.

ELAINE: Of course, there is an idea that islands are small-scale models of the world but in your work you show that the islands that concern you are unique. However, Russell King suggests that we need to be careful how we apply this idea. One of the insights that emerges from the case studies that you present, and that is present in other chapters, is that islands can be heuristic devices. Thérèse, any thoughts about that – does this resonate this from your perspective?

THÉRÈSE: It does. Things to do with islands seem to produce tensions that centre on the constitution of binaries. We may start to apply categories and some of those work but other aspects become obvious because they don't fit, and then you see that you need to 'listen' to what the island has to say – and it may be 'saying' things about that which is internal and external, about things that are consistent or disrupt, unique or in common. My personal experience as a recent arrival from a mainland is that I have become very conscious of water in my day-to-day life: I hear it all the time. It seeps into my consciousness, and the issues and questions that interest me centre more on the water than the land. Water is a particularity of place and a generality of islands. So, you get internal and external influences and perhaps they come together in a unique way that is also held in common with other islands?

ELAINE: And there is, of course, the emergent idea of 'wet ontologies' promulgated by people such as Phil Steinberg and Kim Peters. It is interesting how the focus is not shifting exactly – for archipelagos and water have always been there; but the ways in we are now thinking through them becoming exciting.

JON: And the idea of wet ontologies invites us to think about how islands are becoming interesting for people as sites upon which we are projecting ideas of vulnerability. At the 2015 Exeter conference of the Royal Geographical Society/Institute of British Geographers there was a lot of talk about the world becoming more precarious and uncertain in the Anthropocene, and as much conversation about the ways in which people are projecting these uncertainties and vulnerabilities onto islands, which is another interesting layer and tension within island studies.

ELAINE: Annika, can I ask you about that? Because Kiribati has clearly been declared one of the most vulnerable island groups.

ANNIKA: Yes, that is a complex and complicated area. Goldsmith and Farbotko have both written a fair bit this issue: about how islanders may actively use their agency to portray themselves as vulnerable. They are vulnerable, but it is not just outsiders who are perpetuating such ideas, and that creates tensions on islands, despite the fact that their peoples have often been resilient in the face of, for example, significant climate variability and relative isolation. But anthropogenic climate change is a whole new thing for some of these very vulnerable places. One result is that people are starting to conceptualise the ocean in different ways. I know that Kiribati has experienced really severe flooding in the last few years as a result of various factors including El Niño and sea level rise. I have just seen photos on Facebook [September 2015] where, for the first time, the hospital has been completely flooded out. People are experiencing impacts that they haven't seen before and they are talking about the ocean in fearful ways. I think that adds to the idea that we need to think about seas and oceans always when we think about islands. Hau'ofa and others talk about the sea being a source of regional connection and as sustenance, but it has these multiple meanings now, and possibly always has had different meanings.

ELAINE: David, one of the things that is really interesting about your chapter is the work that you do around the Māori and indigenous understandings of these places. Annika has just made the point that indigenous histories in island places do exemplify high levels of resilience as well, and I'd be interested in hearing your views on this.

DAVID: Quite a few of the New Zealand islands that I have researched have Māori histories, and it is clear that they see islands as connected and not isolated places, so there was and is a lot of mobility from the mainland to islands or from island to island. Europeans don't seem to have that same mind-set about connectivity.

ELAINE: And Russell, one of the questions that you have posed relates to what it means to bracket the phrase 'on its own terms' when we think of studying an island. So let me make the observation that for different disciplines the field is key and field work is a hallmark of what we do and then ask you what did it mean, in practical ways, to study the islands that you've studied?

RUSSELL: When I posed that question I was thinking about the development projects that I have written about, and that Annika's team has also been writing about, but it would apply to any of our chapters – and perhaps to Kate's chapter on seabed mining most of all. At some point, our scholarship has to become practical and the question for me is how do we offer something as island scholars that is leads us to embrace the advances in thinking and practice that grounded field work would provide. I would say it has a lot to do with something that you brought up earlier Elaine: that we are recognising these places as fundamentally different from mainlands and different from one another and, despite that, are being careful not to assume that that difference is more extreme than it actually is.

ELAINE: Have you found it hard to study islands or study on islands?

RUSSELL: I don't know that I have ever conducted longitudinal research on a mainland! So I may not be the best person to answer that question.

ELAINE: Kate, Russell mentioned your seabed project. What has it meant for you in practical ways to study an island, to study on an island, and to study offshore environs?

KATE: Logistically, it was actually quite difficult. In developing my research, I had to consider whether in fact it would be possible for me to go to the field sites I was interested in. It was a practical concern to travel to New Zealand from Arizona and to secure the kind of funding needed. Once there, I had to continue to interrogate the worth and practicality of my methodology – for example, when I was interviewing people or when I was trying to discern who were the key players in seabed mining discourses. At times, I felt real insecurity and discomfort – a marked concern about whether I was influencing people simply by asking the questions I posed. In a place the size of New Zealand many people know each other in particular areas of work and that is true of those involved in seabed mining and governance. So I had to be extra careful to protect my sources of information but was also able to let people guide me in terms of others I could speak with, and that was extremely useful for accessing people working on different elements of the work I was studying.

ELAINE: So, the work you describe is ethically complex, not least because you were dealing with less than six degrees of separation. Marina, did you find comparable dynamics in your studies of the convivial economics of women's cooperatives on Lesvos?

MARINA: On the islands that I have studied you might think you are operating in different worlds, but some people move among islands and some never socialise with other groups. So, in doing my work on Lesvos I could be studying both the women's cooperatives and, in another part of my research on Lesbian enclaves on Eresos, I could live in villages of a thousand people and turn up at events over the course of a year where the different groups I was working with would never intersect. I have found this to be similar in Hawaii, where I am now based: you can move in the Japanese enclave, or among alternative hippy bohemians, all in a relatively small place that still has so much diversity and separation. And I have found it to be the case in the Caribbean, where my work really started: again, I could meet and speak with street vendors and government officials who live in different worlds, and I had access to those as a privileged researcher.

JON: I want to echo what Marina was saying with respect to the Caribbean. I come back to your question and to another you pose about the policy and development implications of island studies. As someone who does participatory research and development I have found it really challenging to frame projects in conditions where people don't necessarily speak to each other and where strong hierarchies may be apparent. I'm thinking of the work I have done with fisher people who decide to drive their own development projects. There are layers upon layers of complications that prevent or delay those initiatives, not least among them Western university systems of education and research, the ways in which development aid works and concentrates in the hands of particular peoples, and the dynamics of island politics as well – and I think that gets more complicated when you are working in archipelagos.

ELAINE: And Annika, did you experience similar dilemmas in your work in Kiribati?

ANNIKA: Yes. Logistically, it is difficult to work in and conduct research on Kiribati because flight schedules are changing all the time – you are getting 'kicked off' flights, for example. You have to be very, very flexible and run with things. In my experience, case study sites changed constantly, and depended on where I could get to and who could come with me. I had concerns about participation: about not wanting to put words in people's mouths; wanting to be honest and transparent about what I was trying to do, and not wanting to say something like 'I am investigating climate change so what do you think has caused these changes in the environment' because they will just say 'climate change'.

On the ground, my approach has been about getting on island time, immersing myself, developing social networks, learning as much of the language as possible, and working through systems established by the churches and mechanisms of traditional governance. I also think flexibility is key. For example, I had planned to use focus groups but they were so unwieldy: the whole community would come together and there would be 40 people in a focus group, which left me dividing people into groups, answering unexpected questions, translating on the go. You will appreciate that it did not work as a method and I needed to be flexible and change my approach.

Coming out of that experience I realise there is what might be called a participatory rhetoric in project documents and among personnel in different agencies but ironically it is just really hard to hear people's voices. I think perhaps that there needs to be more and clearer ways to pass information from communities to governments. And with donors it is just as complex. Many people are too shy to speak, and interactions are set up in ways that mean that donors come away without modifying their ideas at all.

ELAINE: Thank you for that. Now, one of the other questions we have been pondering is whether there a way to use human geography to influence other fields to engage with island studies? And Marina, can I start by asking you what motivated you to ask that question in our early discussions? And in what you ask, is there an implicit assumption that we should be influencing those other fields?

MARINA: Well, remember I am a sociologist, and the spatial turn, while powerful, isn't universal, although it is present in the sub-fields I work in – world-systems theory, for example, and considerations about the organisation of capital and the significant role that islands have played in those dynamics. So, my shift towards human geography is about my saying that space matters when we want to think about social inequalities. For me, it was about asking 'how can we get sociologists to think about islands?' Even though the sociology of development is one of my major fields, I think we remain caught up in the idea that islands are just small places, and that there is nothing particularly critical about how sociological insights are being applied to islands. So that is where it my question was coming from.

RUSSELL: I do remember something in *Island Studies Journal* about these twin issues of smallness and insularity and there was an attempt to determine the contribution of each towards the character of places. In other words, is it the smallness or is it the islandness that makes a place unique or sets it apart? If I remember correctly, the author came down on the side of smallness, and that interests me in general terms in relation to the claims that are made about islands.

ELAINE: Yes that is an interesting one. Does anyone want to respond to that? I think it is an important thread in the conversation that we are having.

THÉRÈSE: In terms of scale, I think that questions of smallness and island-ness are central. I often refer to Bruny Island and Manhattan. On so many axes, they seem incomparable – although I daresay there might be some things that they have in common. My sense is that sometimes we do overstate the idea of islandness. I also look at Tasmania in relation to Bruny and I do not see the same things. And being a mainland person I understand that things which are represented as being about islandness exist in small country towns that may or may not also be remote and isolated. We do need to be careful how we define some of those things as being specifically about islands and not scale – and vice versa.

ANNIKA: That's interesting. I have never really written or generalised any-thing from Kiribati to a wider article about island studies in general. The latter is more of a framework to help me to start thinking about certain characteristics or tensions that might apply to Kiribati, but then I feel the need to go deeper and use ethnography to try and really under-stand Kiribati for itself. So scale, size, these things – they are like tools for thinking.

ELAINE: Kate, is that what motivated you to ask the question 'how can we maintain islands as a productive category while embracing the multiplic-ity and becomings?'

KATE: Yes, and I remain concerned that islands are viewed as some sort of bounded category when, as we have already discussed, we need to under-stand them in multiple ways. In my field work, I travelled to four islands in about as many months: the perspectives were incredibly different and there were many more differences than similarities among those loca-tions. So, I am cautious about using islandness as some sort of explana-tory variable; perhaps it is more of a relationship and a factor, but either way we need to ensure that it is not detrimental to research.

RUSSELL: It is really the claim of causation that is least supported in a lot of island studies. We establish certain findings on the basis of a lot scholar-ship, and then feel tempted to present what we have discovered or con-cluded as if to say 'and this happened on an island – therefore causality'. I think that is best avoided.

THÉRÈSE: Yet, one of the things that I drew from my work is that there is a specific physicality to islands. There is a materiality about bounded places. That boundary is real – even if it shifts. It is there. The ocean is there. And that is a factor in the constitution of particular experience. I agree it is not everything, and it combines with all of the different fac-tors that link into that particular place. The particular mobilities, geo-graphic locations, peoples, and histories are diverse but the island also plays into that. All these other factors will produce something different but the physicality of place, that fact that there are bodies on the ground and they are surrounded by water, surely that has to have an impact and produces specific experiences. There is something quite specific about the physicality of islands that – while it may not be the be-all or a category that can explain everything – is, I think, part of it.

ANNIKA: And that is why islands appeal to human geographers because they elucidate interactions between – on one hand – the physical characteristics of the environment and the places that we create and – on the other hand – relationships between humans and the environment. And that seems definitive of much of human geography.

ELAINE: David, how do you respond to that observation in terms of the questions that you had about connectivity and the loss of islandness? I recall, for example, you thinking about this: if we see the sea as a source of connection, then are there any islands left in the world?

DAVID: Yes. I have seen the sea as a source of connection and that is certainly prevalent in much of the literature but it is possible that a conceptual end-point of that conclusion is that islands no longer exist as such, and I was finding that challenging. I think that we can see the sea as a source of connection, and the journey to an island over water is distinctive, and it marks the island as such, and as different from a mainland. And then there is the work in island studies about bridging effects and that, too, is distinctive. But we need to be careful that 'distinctive' does not become 'deterministic'.

ANNIKA: It is interesting. Hau'ofa's essay on our sea of islands springs to mind here. I find it a really hard question too, David, and Hau'ofa seems to see the Pacific anyway as a region, as Oceania, connected to specific and shared histories of voyages, but obviously he wasn't thinking about that connecting him to the Caribbean or to mainlands.

ELAINE: So, let me ask one final question, then: for each of you, what are the substantive contributions that human geography can make to islands and to island studies, and what, in turn, are the substantive contributions islands and island studies can make to human geography?

KATE: I can address that by building on the last comment about oceans, because it speaks to my background and my studies. In relation to Hau'ofa, I think he was arguing against the ways in which islands have been thought of as constricted, isolated, and separated, and against the ways in which oceans have been seen as really big barriers. Understanding oceans in terms of connections does not erase all distance or difference. It doesn't collapse space to zero. I don't think that is really what Hau'ofa was getting at. All sorts of lands and landscapes may be perceived as barriers – swamps, deserts, and so on, yet these same spaces may be understood in terms of connection, smooth spaces, or other spaces of liberation. Islandness, oceanness, these different forms of landscape allow for certain pursuits and inhibit others. It might be an over-simplification to see ocean as a source of connection just as it is simply to see the category of island as constraining. For me, to include these other ways of thinking is important and should inform human geography as well.

THÉRÈSE: And to suggest that the ocean is a means of connection does not erase the possibility that it is also a barrier. I actually find this tension exciting. It expands our concept of islands and does not erase them. And I think it brings ocean's materiality to bear on the idea of islands, which is something we probably should continue to grapple with and I do find that exciting. In human geography, we think of (just for example) ideas about social justice, development, and world-systems specifically in terms of spatiality and specifically in terms of environment, space, and place, and we advance claims in ways that constitute and are constituted by all of those other things.

ANNIKA: I think that is what Hau'ofa was trying to say about continental discourses that have painted islands as small and remote, and as hopeless actually – economically hopeless. I think part of understanding islands in their own terms involves working to suspect those discourses of conquest and paternalism emanating from the mainlands, discourses where continents are the norm. That invites us to see islands for what they are without prejudice.

MARINA: I want to add a couple of things in conclusion. Part of what Hau'ofa is saying, part of what we are doing in our scholarship and, for those of us who are living and teaching on islands, part of what are we passing on to students about being on islands is about acknowledging the agency of islanders. So, when my students read his work, it is remarkably freeing for many of them in terms of where they can take their thinking about what it means to be on an island. For me as a sociologist, living on an island and being an islander has distinctly shaped my thinking and my social relations; so island matters – and that is something that I have taken from island studies. We can see the social relations and development trajectories shaped by islandness and we can bring that insight back to our disciplines and our practices.

DAVID: I think that this is a really good discussion. Some of the key geographical themes that I have heard us grapple with are place, scale, and location, and there has been the constant presence of our alertness to interdisciplinary approaches as well. I think that is important in geography because it is such a wide subject area and few disciplines are characterised by so many approaches. I also picked up on the idea that islands are paradoxical. There was one such paradox I mentioned earlier – namely prison and paradise, and we have also discussed islands as isolated and connected and as sanctuaries and places of danger or as wild. I think that is something fascinating about the study of islands: that the same phenomenon can be seen in such different ways at the same time.

RUSSELL: I have been thinking about the first half of the question that you posed, which is how does human geography inform or influence island studies? I like the specification of human geography as distinct from other branches of geography such as physical geography – biogeography, for example – because even with uninhabited islands, which we have discussed, we are talking about islands by means of the concepts that we project onto them. I don't know if issues of insularity would be quite so malleable or transformative if islands were simply the domain of geology or biology, or the sole domain of non-human species. I was especially thinking of this when someone brought up the existence of bridges to islands: we debate how that kind of structure changes an island from human perspectives, whereas from biogeographical perspectives the question and answers to it are different. Each of us inserts humanness into our discussions of islands. Then there is the converse of that question: how does island studies influence human geography? I can only speak for myself and my own scholarship, but island studies has exposed me to a literature that I would not have read otherwise. I have read poetry that I never would have been exposed to and I have read theory that has been applied from outside of my field.

JON: Yes, and thinking about the question of what human geography can contribute to island studies and what island studies can contribute back to human geography, I think our discussions about the spatial turn and the associated tropes of relationality, assemblages, archipelagos, and so on are important. These do foreground questions of change and connection underscoring a certain archipelagic turn and archipelagic thinking in island studies, and that foregrounds these tropes of mobilities and movements. And when we think about questions of oceans and water craft and ships, if we think about the Caribbean, or about refugees moving through the Mediterranean, and so on, what is foregrounded are movement and mobility and also questions of rebirth and metamorphosis, and this comes out a lot in island poetry and literature. So I think what island studies can contribute back to the spatial turn relates to these other questions of time and of new understandings of time and new rhythms of life, of island life. Island literature is powerful here, and although that is something that we perhaps did not quite pick up enough in our conversation, from my perspective it is one of the significant things that I have taken from island studies as a relatively new island scholar.

RUSSELL: Agreed, and as a consciously interdisciplinary field, I think island studies has enriched my scholarship in geography. It is interesting to remember that when I was a student I thought island studies should be a sub-field of geography and I have completely turned from that belief and I think it should stand on its own merit.

ELAINE: And that takes us to time, so sincere thanks everyone – lots for us to continue to think about together and in our own work and work with others. Stay well.

10 Retrospect and prospect

Stephen Royle

Island geographies

In 2001, I published a book with a title similar to this one: not *Island Geographies*, rather *A Geography of Islands* (Royle, 2001). It was my first full-length single-authored monograph and thus of considerable importance to my future ennoblement at Queen's University Belfast as the world's only (or hopefully, first) Professor of Island Geography – although Godfrey Baldacchino already then held a broader title of Professor of Island Studies at the University of Prince Edward Island (UPEI), Canada. (He is now Professor of Sociology at the University of Malta, though maintaining links to UPEI.) Maybe the book also played a role in the development of island geography and island studies for it has been often cited and still sells, I am glad to report.

I have been asked to provide some thoughts on island geographies, past, present, and future, and can deal more readily deal with 'retrospect' than 'prospect' given my experience and the perspective of being 'really old' to quote what was said to me recently, somewhat too directly, by an islander on Geomundo, Korea when we exchanged ages. He was 30 and looked 21; I am 65 and obviously look every day of it. The other contributors to this book gave something of their backgrounds in Chapter 9. Regarding my own retrospect, I have had opportunities to detail my journey from urban to island geography, given a Pauline conversion in an abandoned cottage on Dursey Island off the west of Ireland in 1974 when I developed the curiosity to understand the insular decay all around me (Royle, 1999). The most recent such piece was a chapter for one of Baldacchino's many edited books on islands entitled 'Navigating a World of Islands: a 767 Island Odyssey' (Royle, 2015). In *A Geography of Islands*, I had boasted of visiting what now seems a paltry 320 islands; the day before this particular chapter was first written I was on the World Heritage Site island of Yakushima, off Kyushu, which saw my total rise to 843. Some of this island-bagging is shallow self-indulgence, but many of my insular journeys have taken place to conferences or for research over a period when island geography and the broader realm of island studies, to which geography makes such an important contribution, have been developing. 'Prospect' is more speculative; the future is always unknowable and, being so old, presumably I have limited time left to contemplate – never mind contribute to – said prospect.

First, though, let us consider the book's title, *Island Geographies: Essays and Conversations*. 'Essays' are the seven substantive chapters and the editor's introduction. 'Conversations' relate to the innovative two-hour conference call which gathered together the voices and thoughts of the contributors from their locations around the globe, which have been presented as Chapter 9. Then there is 'geography' or rather 'geographies'. Geography is hugely broad, a multidisciplinary subject within its own boundaries. Any person who spends a career in a university geography department (or geography section of a larger unit) will inevitably attend seminars ranging from philosophy at the extreme humanities pole of the subject to geomorphology or meteorology at the scientific edge. A Venn diagram locating traditional academic subjects would have geography close to the centre overlapping with many different fields. This breadth gives geography an important role in synthesising and understanding material from different parts of the academic, indeed the human, endeavour. As a very welcome recent editorial in the British newspaper, *The Guardian*, put it:

> Geography is a subject for our times. It is inherently multidisciplinary in a world that increasingly values people who have the skills needed to work across the physical and social sciences.
>
> (13 August 2015, n.p.)

This book is not directly concerned much with physical geography, although that part of the subject certainly is vital context to the two chapters with New Zealand concerns: Katherine Sammler on seabed mining and David Bade on nature and islands. The main thrust of these two chapters and the others, however, is within human geography, whose key terms as given by Derek Gregory in the *Dictionary of Human Geography* are place, space, and environment.

Place

Two chapters here are based on New Zealand, others are on a Tasmanian island, Kiribati, Lesvos, two Caribbean islands together – St Bartholomew and St Croix – and Barbados. So there is a good range of places. Other books on islands or another iteration of this book, do or could present different sets of islands; island geographers, like all geographers concerned with place, have their study areas. So the range of places within *Island Geographies* in a sense does not matter, for the book does not attempt a comprehensive coverage of subject or place as with Gillespie and Clague's huge *Encyclopedia of Islands* (2009) or Fischer's *Islands: From Atlantis to Zanzibar* (2012). Most readers will be interested in islands and/or island geography generally but if there are any with a particular concern with, say, Kiribati or Lesvos, let them turn to the chapters by Annika Dean, Donna Green, and Patrick Nunn and that by Marina Karides respectively, where they will find much of especial interest to them. Jon Pugh's chapter studies the literature of St Lucia's Nobel Prize-winning author, Derek Walcott, who "wants to develop a language and poetry that praises the Caribbean's unique landscape and culture", a sense of place indeed.

Space

Space concerns distributions and locations where things are happening on the surface of the Earth or in its seas or atmosphere. If 'place' is largely 'where', 'space' can engage with 'what', 'how', also perhaps 'who' and/or 'when' depending on the details of the study. We see here materials on seabed mining, the operation of aid donors, women's co-operatives and cultural heritages. Regarding the meanings of space, at a large scale there is Pugh on Walcott who constitutes 'Caribbean islands as diffracting spaces, composed as multiple and relational trajectories'. By contrast, Thérèse Murray's chapter on the Cape Bruny lighthouse is focused on just one building on an offshore island within the island state of Tasmania, so it is towards the extreme end of geography's scale, although geographies of the body now take the subject into even smaller 'locations'. Regarding space, Murray speaks to the meaning ascribed to that particular lighthouse building located on the boundary between the land and the sea: 'a built expression of attitudes to the ocean, nature and human engagement'.

Environment

'Environment' is a word that like 'geography' itself has a wide range of meanings and attributes. It deals with surroundings, conditions, areas, often with a firm nod towards nature and the natural world, although the origins of the word as given by the *Oxford English Dictionary* relate to fifteenth-century French terms meaning just proximity or surroundings. Environment certainly features in this book. There is the deep sea, whose environment could be ruined by mineral extraction; the human environment experienced by the people of Lesvos with the spatial behaviour of women differing from that of men; the ecology of offshore islands in New Zealand with their opportunities, sometimes contested, for environmental restoration; and the opportunities and pollution problems associated with waste management on two different Caribbean islands.

Thus the book certainly fits within human geography. As to 'islands' – bodies of land surrounded by water – yes again, for the chapters here study islands. Even Sammler's chapter on deep sea mining relates to the spheres of influence and territorial claims to the seabed of islands in the Pacific. Whether the very fact of insularity is significant regarding what happens on and to an island is a matter for debate. Indeed, it is debated in Chapter 9 where Thérèse Murray's take is valuable: 'There is something quite specific about the physicality of islands that – while it may not be the be-all or a category that can explain everything – is, I think, part of it.'

Now let us put together the two words 'island' and 'geographies' forming what the *Oxford English Dictionary* has as a 'compound'. It is a natural pairing in one sense as islands are quintessentially geographical features, defined by their setting within water and any mention of an island sounds echoes of that geography. Be that as it may, 'island geographies' must be a subset of geography restricted in application to the study of these bodies of land surrounded by water. (Let us not complicate matters by engaging with 'island' as a metaphor to mean anything isolated or surrounded, such as oases or isolated mountain tops – sometimes termed 'sky islands'.) So 'island' reduces, or a better word would be focuses, 'geography' in the compound term of the book title. However, given its breadth, as so praised in *The Guardian* editorial, a case might be made that 'geography' does not much reduce the range of research that can be carried out on our bodies of land surrounded by water. The chapters in this book testify to that. Pugh deals with literature in his consideration of Walcott and also display in his treatment of the Barbados Landships; Dean, Green, and Nunn consider international aid and issues to which it gives rise; Karides considers sociology, feminism, also some aspects of economic geography; Murray's take is philosophical; Russell Fielding's study is an essentially practical comparison of waste management issues; Bade deals with ecology and nature and cultural heritage; Sammler, with her deep sea mining, gets into economics and politics and environmental issues. A number of the papers are filtered through a postcolonial lens. This filter applies even to Fielding's waste management for the differences between the positive benefits of the system on St Bartholomew and the dreadful problems of St Croix he ascribes 'to the discrepancies in culture, economics, and political histories that leave these two neighbouring islands so very far apart'.

Retrospect

Where has island geography come from? From a long way back is one glib answer. Ptolemy located the Fortunate Isles (the Canaries) on the western edge of the known world in the second century AD, by which time people had already been pondering about Atlantis for centuries. Later, studies became more focused as with the sometimes stylised depictions of Mediterranean and other islands called *isolarios* which date back to 1420: 'primarily an illustrated guide for travellers in the Aegean and the Levant' (Campbell, 1985, p. 180). Later still, the geography of islands would inform the studies of scientists who would not have called themselves geographers but whose work nonetheless influenced geography among many other fields. Think of Alfred Russel Wallace (1823–1913) with his *Island Life* (1880) or, of course, of Charles Darwin (1809–82). Darwin is associated irretrievably with the Galapagos Islands but spent periods on other islands as well. For example, he was on St Helena from 8 to 14 July 1836. In his *Voyage of the Beagle* (1839), he refused to say anything about Napoleon, instead writing 2,700 words viewing St Helena through a geographical lens: its geology, vegetation, economy, population, fauna, and weather, especially the winds. Darwin (1839, p. 400) wrote about the Galapagos archipelago as a 'little world within itself', which captures neatly one of the tropes of island geography: the theme (and attraction) of the miniature, bounded world. This trope brings up the notion of the island as intensifier: understand the miniature world to help gain an understanding of the larger world; perhaps also the island as laboratory idea, a place where one can test on 'the little world' things that might be applied on a larger scale outside.

Islands continued to fascinate geographers and others, not least because of this otherworldliness captured by Darwin. However, there was always the danger of outsiders depicting islands and island life without proper regard to the people whose homes they were, akin to the troubling 'tourist gaze' described by John Urry (1990). One popular island book I recall from my childhood is *A Pattern of Islands*, published in 1952, and written by Arthur Grimble (1888–1956); it was based on his experiences from 1914 of work-ing for the British colonial authorities in the Gilbert and Ellice Islands (now Kiribati and Tuvalu). Grimble was funny, charming, and self-deprecating. He became a serious scholar of I-Kiribati language and society, and wrote with respect about the people of the islands, but he could not set aside the colonial ethos and era in which he worked. He certainly mocked it, but that did not gainsay the fact that this was the setting for his postings and career, all spent on islands – he had positions in the Seychelles and the Caribbean after leaving the Pacific.

> I was nominated into a cadetship in the Gilbert and Ellice Islands Protectorate at the end of 1913. The cult of the great god Jingo was far from dead. Most English households of the day took it for granted that nobody could be always right, or ever quite right, except an Englishman. The Almighty was beyond doubt Anglo-Saxon and the popular con-ception of Empire resultantly simple. Dominion over palm and pine (or whatever else happened to be noticeably far-flung) was the heaven-conferred privilege of the Bulldog Breed. Kipling had said so. The colo-nial possessions, as everyone so frankly called them, were properties to be administered first and last for the prestige of the lazy little isle where the trumpet orchids blew [i.e. Great Britain]. Kindly administered natu-rally – nobody but the most frightful bounder could possibly question our sincerity about that – but firmly, too, my boy, firmly too, lest the schoolchildren of Empire forget who were the prefects and who the fags [junior pupils].
>
> (Grimble, cited in Lansdown, 2006, pp. 33–4)

What was needed for island scholarship to develop satisfactorily was to move away from prefects and fags, to understand what has been described as 'islands on their own terms', what Grant McCall (1994a), former President of the International Small Island Studies Association, would regard as 'nissology'. Jon Pugh has this in the first couple of lines of his chapter in this book: 'Island scholarship increasingly focuses upon the aim of understanding islands and islanders on their own terms.' He speaks of 'elevating island voices' and cites in support island writers, Walcott (1930–) and the late Epeli Hau'ofa (1939–2009), a Fijian of Tongan descent born in Papua, who features also in Chapter 9's conversations. In terms of my own interests in the small islands off Ireland, it would be a comparison between, for example, *The Aran Islands* (1907) by John Millington Synge (1871–1909) and *The Islandman* (1929) by Tomás O'Crohan (Ó Criomhthain) (1856–1937). Synge was a Dublin playwright (consider, for example, *Playboy of the Western World* or *Riders to the Sea*), a man of letters and learning whose early death at 37 in 1909 cut sadly short a fine literary career. He visited the Aran Islands for several summers around the turn of the twentieth century to better his Irish language skills and to collect material on Irish rural life, which was used in his plays and other writing. He kept journals and from them his book *The Aran Islands* was published. Synge was an outsider, an observer. By contrast, Tomás O'Crohan from Great Blasket Island was an insider, an actor, an islander as affirmed by the title of his autobiography *The Islandman* (An t'Oileánach in its original Irish). Encouraged by a visitor to do so, O'Crohan (1929, p. 244) wrote about his life on Great Blasket Island, its joys and also its sometimes terrible bleakness, conscious of change, 'that the like of us will never be again'. Pugh would 'elevate' O'Crohan's island voice, but this does not mean Synge has no value or that he should be discounted for being an outsider, just that this position must be remembered. In fact, Synge has been praised for his sympathetic yet realistic treatment of the islanders even by islanders. Máirtín Ó Direáin (1910–88), a well-regarded poet from the Aran Islands, compared Synge favourably to the antiquarians who wrote about the many archaeological sites on the Aran Islands for: 'It was not from stone you took your stories,/ but from the wonders in stories by the fire' (Ó Direáin, 1986/2015, n.p.).

Thus, there is now at the very least within island (human) geography and island studies generally an acknowledgement of the importance of studying 'islands on their own terms'. On the front page of islandstudies.ca, home of *Island Studies Journal*, these words appear in bold: 'Island Studies is the global, comparative and inter-disciplinary study of islands on their own terms.' In similar vein, in another now well-established islands journal, *Shima: The International Journal of Research into Island Cultures*, the Editorial Board (2007, n.p.) writes: 'of the principle that external researchers should develop their projects in consultation with island communities and should reciprocate such co-operation with appropriate assistance and facilitation of local cultural initiative'.

The Editorial Board's piece gives a history of island studies back to what is now counted as the first 'Islands of the World' conference on Vancouver Island in 1986, and shows how that series developed and became run by the International Small Island Studies Association, which – along with RETI (Réseaux d'Excellence des Territoires Insulaires, the network of island universities) – has *Island Studies Journal* as its house outlet. A number of other island bodies and journals are mentioned, including *Shima* itself, established by Philip Hayward, who also set up the Small Island Cultures Research Initiative, which also runs an annual 'International Small Islands Cultures' conference.

Geographers are well represented at these multidisciplinary conferences, and within the subject itself, the International Geographical Union has its Commission on Islands (http://igu-islands.giee.ntnu.edu.tw). Thus there are conferences galore, a number of specialist journal outlets, and many edited books on island studies, including this one related specifically to island geography. Scholars also publish about islands in other journals.

A retrospective piece in *Island Studies Journal* was written by Russell King (2015). King had been commissioned in 1991 by the Institute of Island Studies at UPEI, to consider the potential for a journal focusing on island studies. Such a publication, King stressed, would have to be high quality and rigorously refereed, international in scope, and multidisciplinary in character. Based 'more on intuition than anything else', King (2015, p. 9) felt at the time 'that there is a market for such a journal, and that the journal will be an attractive and exciting venture'. *Island Studies Journal* did not actually appear until 2006 but that apart, King was satisfied that it had fulfilled his original criteria. In similar vein, I recall vividly at the 'Islands of the World XI' conference on Bornholm, Denmark, in 2010, a young literature scholar remarking in passing something to the effect that 'in island studies we ...', thus completely accepting the fact that this discipline existed. Earlier there had been speculation 'fleshing out the productively contested faultlines and parameters of the field' (Stratford, 2015, p. 143). Appositely that quote is from Elaine Stratford, editor of this book, in a review of ten years of *Island Studies Journal*, where her analysis notes that geographers are by far the most common contributors to the journal.

Prospect

In her 2015 paper, Stratford went on to suggest ways in which *Island Studies Journal* and the authors therein could improve visibility in the increasingly competitive world of academic scholarship. (One of the great benefits of being 'really old' is not having to compete anymore!) Among a number of invited comments on King's 2015 reflections was one by Lisa Fletcher (2015) who works within what she had earlier described as 'island literary studies' at the intersection between literature and geography (Fletcher, 2011, p. 7). She would like to have seen more literary studies in the journal.

Other commentators would like to have seen more contributions from Asia (Tsai, 2015) and also from the developing world – there having been no contributions to *Island Studies Journal* from Africa (Nurse, 2015). Technological developments in theory should make it ever easier for people from poorer countries to play a greater role in publishing: 'This is increasingly achievable given modern communication technologies and the globalization of higher education' (Nurse, 2015, p. 14). This observation, of course, pertains to islanders themselves. However, for all the talk of 'islands on their own terms' and privileging the island voice, publications will probably continue be dominated by those from the usual rich countries in Europe, North America, and Oceania. That there will continue to be publications on islands I do not doubt, for island geography/studies is established now to the extent that there are a number of postgraduate courses in the field at universities across the globe. Some of these students will, one hopes, move on through being early career researchers into fully-fledged scholars. For people who become practitioners in (or on) islands, communicating their experiences and ideas in some form is to be anticipated and welcomed. There are a number of specialist institutions as well: after all, I write this from the Kagoshima University Research Center for the Pacific Islands, publisher of *South Pacific Studies*. This outlet, into Volume 35 now, more than some others does publish studies on and/or with authors from some of the island nations not well represented elsewhere. Recent issues have published papers on Papua New Guinea, Maluku (Indonesia), Mindanao (Philippines), The Federated States of Micronesia, and Vanuatu, as well as on Japanese islands. Authors have come from Indonesia, Malaysia, the Philippines, and Vietnam, as well as Japan. *The Journal of Marine and Island Cultures*, into its fourth year, edited by Sun Kee Hong of Mokpo University, Korea, also goes some way to tackle the Asia deficit elsewhere and has published articles regarding islands in Indonesia, Japan, Korea, the Philippines, Singapore, Sri Lanka, and Taiwan. In sum, one might expect to see a greater spread of places and authors within island scholarship, although higher income nations will inevitably continue to predominate.

As to future topics, that is harder to predict. The 'islands on their own terms' idea is well-established – thus it plays a considerable role in this book's 'conversations' in the last chapter – so that there may well be greater consideration not just of the island voice, but along the lines of the phrase there used in Chapter 9 by Annika Dean: 'part of understanding islands in their own terms involves working to suspect those discourses of conquest and paternalism emanating from the mainlands, discourses where continents are the norm'.

Recent developments in island linkages, Hau'ofa's sea of islands (1994), including but moving beyond archipelagos might be expected to continue (for example, see Baldacchino, 2015), as might work related to the idea of the 'aquapelago': 'the manner in which the aquatic spaces between and around groups of islands [is] utilised and navigated in a manner that is fundamentally interconnected with and essential to social groups' habitation of land' (Hayward, 2012, p. 1).

Already Christian Fleury has written of the 'island/sea/territory relationship' (2013) and there is some discussion in Chapter 9 about 'wet ontologies' and about 'the sea as a source of connection'. This may be in 'prospect' but the sea was not always the barrier it seemed to become as modern transportation systems developed; it used to be the sea that was the highway.

What seems certain is that the fragilities (the previous chapter has 'vulnerabilities') of the island realm will stimulate interest, research, and publications. Baldacchino has already edited *Extreme Heritage Management: The Practices and Policies of Densely Populated Islands* (2012) in which I had the opportunity to consider islands as miners' canaries potentially warning of more general impending doom (Royle, 2012). At greater length and depth is John Connell's *Islands at Risk* (2013). The impacts of climate change, especially sea level rise, is a massive part of such risk and seems sure to come within the purview of island scholars, be they geographers or from other disciplines. Another hot topic might be migration, now troubling many islands, such as those on the fringes of Europe. Other established topics will doubtless continue, including island tourism, but the once ready word association between 'island' and 'paradise' may be more difficult to sustain in what seems set to be an uncertain future.

Bibliography

Agardy, T., 2010. *Ocean Zoning: Making Marine Management More Effective*. Washington, DC: Earthscan.

Alderman, D., 2003. Street names and the scaling of memory: the politics of commemorating Martin Luther King, Jr within the African American community. *Area*, 35(2), pp. 163–173.

Allard, J. and Matthaei, J., 2008. Introduction. In: J. Allard, C. Davidson, and J. Matthaei, eds, *Solidarity Economy: Building Alternative for People and Planet: Papers and Reports from the 2007 US Social Forum*. Chicago, IL: ChangeMaker Publications, pp. 1–18.

Allen, M.G., 2015. Framing food security in the Pacific Islands: empirical evidence from an island in the Western Pacific. *Regional Environmental Change*, 15(7), pp. 1341–1353. Available at: http://link.springer.com/10.1007/s10113-014-0734-5

Allsopp, R., 1972. The question of Barbadian culture (No. 1). *Bajan Magazine*.

Anderson, A., 2002. A fragile plenty: pre-European Māori and the New Zealand environment. In: E. Pawson and T. Brooking, eds, *Environmental Histories of New Zealand*. Auckland: Oxford University Press, pp. 19–34.

Anderson, K., 1995. Culture and nature at the Adelaide Zoo: at the frontiers of 'human' geography. *Transactions of the Institute of British Geographers*, 20(3), pp. 275–294.

Anonymous, 2009. Island calls for better recycling habits. *St Barth Weekly*, 182, 4 December, p. 2.

Anonymous, 2012. Fire breaks out at Anguilla landfill. *St Croix Source*, 22 June. Available at: http://stcroixsource.com/content/news/local-news/2012/06/22/fire-breaks-out-anguilla-landfill [Accessed 31 December 2014].

Arnold, R., 1994. *New Zealand's Burning: The Settler World in the Mid-1880s*. Wellington: Victoria University Press.

Atkinson, J., 2010. China–Taiwan diplomatic competition and the Pacific Islands. *The Pacific Review*, 23(4), pp. 407–427.

Auckland Star, 1935. Rangitoto: Protest against invasion. *Auckland Star*, 5 April, p. 6.

Bade, D., 2010. Issues and tensions in island heritage management: a case study of Motuihe Island, New Zealand. *Island Studies Journal*, 5(1), pp. 25–42.

Bade, D., 2013. Managing cultural heritage in "natural" protected areas: case studies of Rangitoto and Motutapu islands in Auckland's Hauraki Gulf. Unpublished PhD thesis, University of Auckland.

Baird, M., 2013. 'The breath of the mountain is my heart': indigenous cultural landscapes and the politics of heritage. *International Journal of Heritage Studies*, 19(4), pp. 327–340.

Baldacchino, G., 2004. The coming of age of island studies. *Tijdschrift voor Economische en Sociale Geografie*, 95(3), pp. 272–283.

Baldacchino, G., 2005. Islands: objects of representation. *Geografiska Annaler Series B, Human Geography*, 87(4), pp. 247–251.

Baldacchino, G., 2006a. Islands, island studies, Island Studies Journal. *Island Studies Journal*, 1(1), pp. 3–18.

Baldacchino, G., 2006b. Managing the hinterland beyond: two ideal-type strategies of economic development for small island territories. *Asia Pacific Viewpoint*, 47(1), pp. 45–60.

Baldacchino, G., 2008. Studying islands: on whose terms? Some epistemological and methodological challenges to the pursuit of island studies. *Island Studies Journal*, 3(1), pp. 37–56.

Baldacchino, G., 2010. *Island Enclaves: Offshoring Strategies, Creative Governance, and Subnational Island Jurisdictions*. Montreal: McGill-Queen's University Press.

Baldacchino, G., ed., 2007. *A World of Islands: An Island Studies Reader*. Charlottetown, PEI: Institute of Island Studies.

Baldacchino, G., ed., 2012. *Extreme Heritage Management: The Practices and Policies of Densely Populated Islands*. Oxford: Berghahn Books.

Baldacchino, G., ed., 2015. *Archipelago Tourism: Policies and Practices*. Farnham: Ashgate.

Baldacchino, G. and Royle, S.A., 2010. Postcolonialism and islands: introduction. *Space and Culture*, 13(2), pp. 140–143.

Ballara, A., 2010. Tenetahi, Wiremu Te Heru, biography. *Dictionary of New Zealand Biography*. Available at: www.TeAra.govt.nz/en/biographies/2t36/1 [Accessed 31 January 2015].

Bankoff, G., 2001. Rendering the world unsafe: "vulnerability" as Western discourse. *Disasters*, 25(1), pp. 19–35.

Barnett, J., 2005. Titanic states? Impacts and responses to climate change in the Pacific Islands. *Journal of International Affairs*, 59(1), pp. 203–219.

Barnett, J. and Adger, W.N., 2003. Climate dangers and atoll countries. *Climatic Change*, 61(3), pp. 321–337.

Barnett, J. and Campbell, J., 2010. *Climate Change and Small Island States: Power, Knowledge and the South Pacific*. London: Earthscan Publications.

Barney, K., 2007. On the margins: Torres Strait islander women performing contemporary music. *Shima: The International Journal of Research into Island Cultures*, 1(2), pp. 70–90.

Belich, J., 2001. *Paradise Reforged: a History of the New Zealanders from the 1880s to the Year 2000*. Auckland: Penguin Press.

Bell, C., 1996. *Inventing New Zealand: Everyday Myths of Pākehā Identity*. Auckland: Penguin Books.

Bellingham, P., Towns, D., Cameron, E., Davis, J., Wardle, D., Wilmshurst, J. and Mulder, C., 2010. New Zealand island restoration: seabirds, predators, and the importance of history. *New Zealand Journal of Ecology*, 34(1), pp. 115–136.

Bertram, G., 1999. Economy. In: M. Rapaport, ed., *The Pacific Islands: Economy and Society*. Hawaii: Bess Press, pp. 337–350.

Bertram, G. and Watters, R.F., 1985. The MIRAB economy in South Pacific Island States. *Pacific Viewpoint*, 26(3), pp. 497–519.

Bertram, G. and Watters, R.F., 1986. The MIRAB process: earlier analysis in context. *Pacific Viewpoint*, 27(1), pp. 47–59.

Besio, K., Johnston, L. and Longhurst, R., 2008. Sexy beasts and devoted mums: narrating nature through dolphin tourism. *Environment and Planning A*, 40(5), pp. 1219–1234.

Best, C., 2001. *Roots to Popular Culture: Barbadian Aesthetics: Kamau Brathwaite to Hard-core Styles*. London: Macmillan.

Bhabha, H. 1994. *The Location of Culture*. London: Routledge.

Blum, H., 2013. Introduction: oceanic studies. *Atlantic Studies*, 10(2), pp. 151–155.

BoM & CSIRO [Australian Bureau of Meteorology and the Commonwealth Scientific and Industrial Research Organisation], 2011. *Kiribati. Climate Change in the Pacific: Scientific Assessment and New Research. Volume 2: Country Reports*. Canberra: Pacific Climate Change Science Program, AusAid.

Borooah, V.K., 2014. *Europe in an Age of Austerity*. London: Palgrave Macmillan.

Boschen, R., Rowden, A., Clark, M. and Gardner, J., 2013. Mining of deep-sea seafloor massive sulfides: a review of the deposits, their benthic communities, impacts from mining, regulatory frameworks and management strategies. *Ocean and Coastal Management*, 84, pp. 54–67.

Bourdieu, P., 1998. *Acts of Resistance: Against the Tyranny of the Market*. New York: New Press.

Bourdin, G., 2012. *History of St Barthélemy*. W.A. von Mueffling.

Brassey, R., 1993. *Demolition of Baches, Rangitoto [Robert Brassey correspondence with Jim Henry]*. Auckland: Department of Conservation.

Brathwaite, E.K., 1967. *The Arrivants: A New World Trilogy*. Oxford: Oxford University Press.

Brathwaite, E.K., 1971. *The Development of Creole Society in Jamaica, 1770–1820*. Oxford: Clarendon Press.

Brathwaite, E.K., 1999. *Conversations with Nathaniel Mackey*. Staten Island, NY: We Press.

Briguglio, L., 1995. Small island developing states and their economic vulnerabilities. *World Development*, 23(9), pp. 1615–1632.

Brooking, T. and Pawson, E., eds, 2011. *Seeds of Empire: The Environmental Transformation of New Zealand*. London: I.B. Tauris.

Brooking, T., Hodge, R. and Wood, V., 2002. The grasslands revolution reconsidered. In: E. Pawson and T. Brooking, eds, *Environmental Histories of New Zealand*. Auckland: Oxford University Press.

Bührs, T. and Bartlett, R., 1993. *Environmental Policy in New Zealand: The Politics of Clean and Green?* Auckland: Oxford University Press.

Burnett, C.D., 2005. The edges of empire and the limits of sovereignty: American guano islands. *American Quarterly*, 57(3), pp. 779–803.

Burrowes, M., 2005. The cloaking of a heritage: The Barbados Landship. In: G. Heuman and D. Trotman, eds, *Contesting Freedom: Control and Resistance in the Post-Emancipation Caribbean*. Oxford: Macmillan Caribbean.

Butler, D., Lindsay, T. and Hunt, J., 2014. *Paradise Saved: The Remarkable Story of New Zealand's Wildlife Sanctuaries and how they are Stemming the Tide of Extinction*. Auckland: Random House.

Campbell, E.M.J., 1985. Material of nautical cartography from c.1550–1650 in the Bodleian Library, Oxford. *Revista da Universidade de Coimbra*, 32, pp. 179–185.

Carter, J., 2010. Displacing Indigenous cultural landscapes: the naturalistic gaze at Fraser Island World Heritage Area. *Geographical Research*, 48(4), pp. 398–410.

Casey, E.S. 1996. How to get from space to place in a fairly short stretch of time: phenomenological prolegomena. In: S. Feld and K.H. Basso, eds, *Senses of Place*. Santa Fe, NM: School of American Research Press, pp. 13–52.

Cavell, S., 1996. *A Pitch of Philosophy*. Cambridge, MA: Harvard University Press.

Charmaz, K., 2006. *Constructing Grounded Theory: A Practical Guide through Qualitative Analysis*. Thousand Oaks, CA: SAGE Publications.

Chatham Rock Phosphate Limited, 2014. Application for a marine consent to undertake a discretionary activity. NZ Environmental Protection Authority. Available at: www.epa.govt.nz/eez/EEZ000006/EEZ000006CRPprescribedapplicationform.pdf [Accessed 28 July 2014].

Choo, H.Y. and Freree, M., 2010. Practicing intersectionality in sociological research: a critical analysis of inclusions, interactions, and institutions in the study of inequalities. *Sociological Theory*, 28(2), pp. 129–149.

Clark, E. and Tsai, H-M., 2012. Islands: ecologically unequal exchange and landesque capital. In: A. Hornborg, B. Clark and K. Hermele, eds, *Ecology and Power: Struggles over Land and Material Resources in the Past, Present and Future*. London and New York: Routledge, pp. 52–67.

Clarke, R., 2001. Roots: a genealogy of the 'Barbadian Personality'. In: D. Marshall and G.D. Howe, eds, *The Empowering Impulse: The Nationalist Tradition of Barbados*. Kingston, Jamaica: Canoe Press, pp. 301–349.

Clayton, D., 2011. Subaltern space. In: J.A. Agnew and D.N. Livingston, eds, *The SAGE Handbook of Geographical Knowledge*. Thousand Oaks, CA: SAGE Publications, p. 246.

Collins, J., 2003. FAA gets tough with USVI over St Croix airport. *Caribbean Business*, 14 August, p. 38.

Collins, P.H., 1991. *Black Feminist Thought: Knowledge, Consciousness, and the Politics of Empowerment*. London and New York: Routledge.

Collymore, F.A., 1955/1992. *Barbadian Dialect*, 6th edition. Barbados: Barbados National Trust.

Connell, J., 2013. *Islands at Risk? Environments, Economies and Contemporary Change*. Cheltenham: Edward Elgar Publishing.

Connell, J. and Lea, J., 1992. 'My country will not be there': global warming, development and the planning response in small island states. *Cities*, 9(4), pp. 295–309.

Cooke, B. and Kothari, U., 2001. *Participation: The New Tyranny?* London and New York: Zed Books.

Corbett, J. and Connell, J., 2015. All the world is a stage: global governance, human resources, and the 'problem' of smallness. *The Pacific Review*, 28(3), pp. 1–25.

Cotis, J-P., 2011. *Recensement de la Population*. Paris: INSEE.

Cottrell, N., 1984. *A History and Bibliography of Motutapu and Rangitoto Islands*. Auckland: Department of Lands and Survey.

Cousin, B. and Chauvin, S., 2013. Islanders, immigrants and millionaires: the dynamics of upper-class segregation in St Barts, French West Indies. In: I. Hay, ed., *Geographies of the Super-Rich*. Cheltenham: Edward Elgar Publishing, pp. 186–200.

Cresswell, T., 2004. *Place: A Short Introduction*. Oxford: Blackwell Publishers.

Cronon, W., 1995. The trouble with wilderness; or, getting back to the wrong nature. In: W. Cronon, ed., *Uncommon Ground: Rethinking the Human Place in Nature*. New York: W.W. Norton and Company, pp. 69–90.

Cronon, W., 1996. The trouble with wilderness: or, getting back to the wrong nature. *Environmental History*, 1(1), pp. 7–28.

Cronon, W., 2003. The riddle of the Apostle Islands. *Orion*, 22(May/June), pp. 36–42.

Crosby, A., 1986. *The Biological Expansion of Europe, 900–1900*. Cambridge: Cambridge University Press.

Crowards, T., 2002. Defining the category of 'small' states. *Journal of International Development*, 14(2), pp. 143–179.

Crown Minerals Act 1991. [online] Available at: www.legislation.govt.nz/act/public/1991/0070/latest/DLM242536.html [Accessed 12 March 2015].

Crown Minerals Amendment Act 2013. [online] Available at: www.legislation.govt.nz/act/public/2013/0014/latest/DLM4756113.html [Accessed 12 March 2015].

Darwin, C., 1839. *Narrative of the Surveying Voyages of His Majesty's Ships Adventure and Beagle Between the Years 1826 and 1836*. London: Henry Colburn.

Daugherty, C., Towns, D., Atkinson, I. and Gibbs, G., 1990. The significance of the biological resources of New Zealand islands for biological restoration. In: D. Towns, I. Atkinson and C. Daugherty, eds, *Ecological Restoration of New Zealand Islands*. Wellington: Department of Conservation, pp. 9–21.

Davidson, J., 1990. Key archaeological features of the offshore islands of New Zealand. In: D. Towns, I. Atkinson and C. Daugherty, eds, *Ecological Restoration of New Zealand Islands*. Wellington: Department of Conservation, pp. 150–155.

de Albuquerque, K. and McElroy, J.L., 1999. Correlates of race, ethnicity and national origin in the United States Virgin Islands. *Social and Economic Studies*, 48(3), pp. 1–42.

De Certeau, M., 1984. *The Practice of Everyday Life*. Berkeley: University of California Press.

De Leo, F.C., Smith, C.R., Rowden, A.A., Bowden, D.A. and Clark, M.R., 2010. Submarine canyons: hotspots of benthic biomass and productivity in the deep sea. *Proceedings of the Royal Society B: Biological Sciences*, rspb20100462.

Deep Sea Mining Campaign, 2013. Campaign overview. Available at: www.deepseaminingoutofourdepth.org/about/ [Accessed 12 March 2015].

Deepwater Group Limited, 2014. Call for seabed 'legal anomalies' to be resolved. Press release 19 May 2014. Available at: http://deepwatergroup.org/dwg-release-call-to-for-seabed-legal-anomalies-to-be-resolved [Accessed 12 March 2015].

Deleuze, G. and Guattari, F., 1983. *Anti-Oedipus: Capitalism and Schizophrenia*. Minneapolis: University of Minnesota Press.

Deleuze, G. and Guattari, F., 1987. *A Thousand Plateaus. Capitalism and Schizophrenia* (with notes on the translation and acknowledgements by Brian Massumi). Minneapolis: University of Minnesota Press.

DeLoughrey, E., 2001. 'The litany of islands, the rosary of archipelagoes': Caribbean and Pacific archipelagraphy. *Ariel: A Review of International English Literature*, 32(1), pp. 21–51.

DeLoughrey, E.M., 2007. *Routes and Roots: Navigating Caribbean and Pacific Island Literatures*. Honolulu: University of Hawaii Press.

Dening, G., 2004. Deep times, deep spaces. In: B. Klein and G. Mackenthun, eds, *Sea Changes: Historicizing the Ocean*. London and New York: Routledge, pp. 13–36.

Department of Conservation, 1992. *Motutapu: Brochure Commemorating the Visit of the Duke of Edinburgh and International President, World Wide Fund for Nature*, Auckland: DOC.

Department of Conservation, no date. Offshore Islands and Conservation. Available at: www.doc.govt.nz/conservation/land-and-freshwater/offshore-islands/new-zealands-offshore-islands/ [Accessed 31 January 2015].

Depraetere, C., 2008. The challenge of nissology: a global outlook on the world archipelago part II: The global and scientific vocation of nissology. *Island Studies Journal*, 3(1), pp. 17–36.

Diamond, J., 1990. New Zealand as an archipelago: an international perspective. In: D. Towns, I. Atkinson and C. Daugherty, eds, *Ecological Restoration of New Zealand Islands*. Wellington: Department of Conservation, pp. 3–8.

Díaz, M.E., 2011. Cost–benefit analysis of a waste to energy plant for Montevideo; and waste to energy in small islands. Unpublished MSc thesis, Columbia University, New York.

Dickens, P., 2009. The cosmos as capitalism's outside. *The Sociological Review*, 57(s1), pp. 66–82.

Dingler, J., 2005. The discursive nature of nature: towards a post-modern concept of nature. *Journal of Environmental Policy and Planning*, 7(3), pp. 209–225.

Doan, P.L., 2010. The tyranny of gendered space – reflections from beyond the gender dichotomy. *Gender, Place, and Culture: A Journal of Feminist Geography*, 17(5), pp. 635–654.

Dodd, A., 2007. Management of the Motutapu archaeological landscape. *Archaeology in New Zealand*, 50(4), pp. 253–274.

Donner, S.D. and Webber, S., 2014. Obstacles to climate change adaptation decisions: a case study of sea-level rise and coastal protection measures in Kiribati. *Sustainability Science*, 9(3), pp. 331–345.

Dookhan, I., 1994. *A History of the Virgin Islands of the United States*. Kingston, Jamaica: Canoe Press.

Douglas, A., 2010. Between a rock and a hard place: does the Treaty of Waitangi provide an avenue for iwi to assert legal interests in minerals in the Crown owned conservation estate? Unpublished PhD thesis, University of Otago, Dunedin.

Downes, A., 2000. Sailing from colonial into national waters: a history of the Barbados Landship. *Journal of the Barbados Museum and Historical Society*, 46, pp. 93–122.

Dummer, T.J., Cook, I.G., Parker, S.L., Barrett, G.A. and Hull, A.P., 2008. Promoting and assessing 'deep learning' in geography fieldwork: an evaluation of reflective field diaries. *Journal of Geography in Higher Education*, 32(3), pp. 459–479.

Eade, D., 2007. Capacity building: who builds whose capacity? *Development in Practice*, 17(4/5), pp. 630–639.

Eden, S., Tunstall, S. and Tapsell, S., 1999. Environmental restoration: environmental management or environmental threat? *Area*, 32(2), pp. 151–159.

Edge, K., 2004. *Secretary Island Restoration Project*. Invercargill: Department of Conservation.

EEZ Act (Exclusive Economic Zone and Continental Shelf (Environmental Effects)) 2012. Available at: www.legislation.govt.nz/act/public/2012/0072/latest/DLM3955428.html [Accessed 12 March 2015].

EEZ Amendment Act 2013. Available at: www.legislation.govt.nz/act/public/2013/0085/latest/DLM5644760.html?src=qs [Accessed 12 March 2015].

Elden, S. 2007. Government, calculation, territory. *Environment and Planning D: Society and Space*, 25(3), pp. 562–580.

Elden, S., 2010. Land, terrain, territory. *Progress in Human Geography*, 34(6), pp. 799–817.

Eldrege, C., 1991. *Pacific Parallels: Artists and the Landscape in New Zealand.* Washington, DC: New Zealand–United States Arts Foundation.

Emerson, R.W., 2015. *Self-Reliance and Other Essays.* New York: Seedbox Press.

Environmental Protection Authority, 2014. Trans-Tasman Resources Ltd marine consent decision. Available at: www.epa.govt.nz/EEZ/EEZ000004/TransTasmanResourcesdecision17June2014.pdf [Accessed 28 April 2015].

Environmental Protection Authority, 2015. Decision on marine consent application by Chatham Rock Phosphate Limited to mine phosphorite nodules on the Chatham Rise. Available at: www.epa.govt.nz/eez/EEZ000006/EEZ000006CRP%20Final%20Version%20of%20Decision.pdf [Accessed 28 April 2015].

Environmental Protection Authority, n.d. The Exclusive Economic Zone and Continental Shelf. Available at: www.epa.govt.nz/EEZ [Accessed 2 May 2014].

European Social Commission, 2011. Small and medium sized enterprises: the social economy. Available at: http://ec.europa.eu/enterprise/policies/sme/promoting-entrepreneurship/social-economy/ [Accessed 15 August 2011].

Falkland, A. and White, I., 2009. Management of freshwater lenses on small Pacific islands. *Hydrogeology Journal*, 18(1), pp. 227–246.

Fankhauser, S. and Burton, I., 2011. Spending adaptation money wisely. *Climate Policy*, 11(3), pp. 1037–1049.

Fergusson-Jacobs, N., 2013. *Full Steam Ahead: Locating the Barbados Landship.* Barbados: Landship Black Star Productions.

Ferrier, T., Hepper, J. and Wells, M., 2011/2014. *Bruny Island Tourism Strategy*, revised edition. Hobart: Kingborough Council.

Fielding, R. and Mathewson, K., 2012. Queen of the Caribbees: farming and fishing foci on the island of Nevis. *Focus on Geography*, 55(4), pp. 132–139.

Fischer, S.R., 2012. *Islands: From Atlantis to Zanzibar.* London: Reaktion Books.

Fisheries (Benthic Protection Areas) Regulations, 2007. Available at: www.legislation.govt.nz/regulation/public/2007/0308/latest/whole.html#DLM973975 [Accessed 28 December 2015].

Fisheries Act 1996. Available at: www.legislation.govt.nz/act/public/1996/0088/latest/DLM394192.html [Accessed 28 December 2015].

Flannery, T., 1994. *The Future Eaters: An Ecological History of the Australasian Lands and People.* Sydney: Reed Books.

Fletcher, L., 2011a. '... some distance to go': a critical survey of Island Studies. *New Literatures Review*, 47/48, pp. 17–34.

Fletcher, L., 2011b. Reading the postcolonial island into Amitav Ghosh's *The Hungry Tide*. *Island Studies Journal*, 6(1), pp. 3–16.

Fletcher, L., 2015. Island (literary) studies. *Island Studies Journal*, 10(1), pp. 11–12.

Fleury, C., 2013. The island/sea/territory relationship: towards a broader and three dimensional view of the aquapelagic assemblage. *Shima: The International Journal of Research into Island Cultures*, 7(1), pp. 1–13.

Freeman, C., 2000. *High Tech and High Heel in the Global Economy: Women, Work, and Pink-Collar Identities.* Durham, NC: Duke University Press.

Freire, P., 2005. *Pedagogy of the Oppressed.* New York and London: Continuum.

Galbreath, R., 2002. *Scholars and Gentlemen Both: G.M. and Allan Thomson in New Zealand Science and Education.* Wellington: The Royal Society of New Zealand.

Gardener, M., 1993. Eruption on Motutapu. *New Zealand Listener*, 22 May, pp. 34–37.

Geertz, C., 1963. *Peddlers and Princes: Social Development and Economic Change in Two Indonesian Towns.* Chicago, IL: University of Chicago Press.

Gentry, K., 2006. Place, heritage and identity. In: K. Gentry and G. McLean, eds, *Heartlands: New Zealand Historians Write About Places Where History Happened.* Auckland: Penguin Books, pp. 13–26.

George, S., 2010. *Whose Crises, Whose Future? Towards a Greener, Fairer, Richer World.* Cambridge: Polity Press.

Gibson-Graham, J.K., 2006. *The End of Capitalism (As We Knew It): A Feminist Critique of Political Economy.* Oxford: Blackwell Publishers.

Gillespie, R. and Clague, D., 2009. *Encyclopedia of Islands.* Berkeley: University of California Press.

Gillis, J., 2003. Taking history offshore: Atlantic islands in European minds. In: R. Edmond and V. Smith, eds, *Islands in History and Representation.* London and New York: Routledge, pp. 19–31.

Gillis, J., 2004. *Islands of the Mind: How the Human Imagination Created the Atlantic World.* New York: Palgrave Macmillan.

Ginn, F., 2008. Extension, subversion, containment: eco-nationalism and (post) colonial nature in Aotearoa New Zealand. *Transactions of the Institute of British Geographers*, 33(3), pp. 335–353.

Glasby, G.P., 2000. Lessons learned from deep-sea mining. *Science*, 289(5479), pp. 551–553.

Glaser, B.G. and Strauss, A.L., 1967. *The Discovery of Grounded Theory: Strategies for Qualitative Research.* New York: Aldine de Gruyter.

Glassner, M.I., 1991. The frontiers of Earth – and of political geography: the sea, Antarctica and outer space. *Political Geography Quarterly*, 10(4), pp. 422–443.

Goldsmith, M., 2015. The big smallness of Tuvalu. *Global Environment*, 8(1), pp. 134–151.

Government of Kiribati, 2012. *Kiribati Development Plan 2012–2015.* Tarawa, Kiribati.

Government of Kiribati, 2015. Relocation. Available at: www.climate.gov.ki/category/action/relocation/ [Accessed 22 May 2015].

Grace-McCaskey, C., 2012. Fishermen, politics, and participation: an ethnographic examination of commercial fisheries management in St Croix, US Virgin Islands. Unpublished PhD thesis, University of South Florida, Tampa.

Graham, B., Ashworth, G. and Tunbridge, J., 2000. *A Geography of Heritage: Power, Culture and Economy.* New York: Arnold Publishing.

Greenpeace International, 2014. Deep sea mining. Available at: www.greenpeace.org/international/en/campaigns/oceans/marine-reserves/deep-sea-mining/ [Accessed 20 March 2014].

Gregory, D., 2009. Human geography. In: D. Gregory, R.J, Johnston, G. Pratt, M.J. Watts and S. Whatmore, eds, *The Dictionary of Human Geography.* Oxford: Blackwell Publishers, pp. 350–354.

Gregory, M.R., 2009. Environmental implications of plastic debris in marine settings – entanglement, ingestion, smothering, hangers-on, hitch-hiking and alien invasions. *Philosophical Transactions of the Royal Society B: Biological Sciences*, 364(1526), pp. 2013–2025.

Griffin, C.E., 1997. *Democracy and Neoliberalism in the Developing World: Lessons from the Anglophone Caribbean.* Aldershot: Ashgate.

Griffiths, T., 1996. History and natural history. In: T. Griffiths, *Hunters and Collectors: The Antiquarian Imagination in Australia.* Melbourne: Cambridge University Press.

Grimble, A., 1952. *A Pattern of Islands*. London: John Murray.

Groenewald, T., 2004. A phenomenological research design illustrated. *International Journal of Qualitative Methods*, 3(1), pp. 1–26.

Grydehøj, A., 2014. Constructing a centre on the periphery: urbanization and urban design in the island city of Nuuk, Greenland. *Island Studies Journal*, 9(2), pp. 205–222.

Grydehøj, A., Pinya, X., Cooke, G., Dorath, N., Elewa, A., Kelman, I., Pugh, J., Schick, L. and Swaminathan, R., 2015. Returning from the horizon: introducing urban island studies. *Urban Island Studies*, 1(1), pp. 1–19.

Hacker, S., 1989. *Pleasure, Power and Technology: Some Tales of Gender, Engineering and the Co-operative Workplace*. London and New York: Routledge.

Haggerty, J. and Campbell, H., 2009. Farming and the environment. Available at: www.teara.govt.nz/en/farming-and-the-environment [Accessed 31 January 2015].

Hanson, S. and Pratt, G., 1995. *Gender, Work, and Space*. London and New York: Routledge.

Haque, T., Knight, D. and Jayasuriya, D., 2015. Capacity constraints and public financial management in small Pacific island countries. *Asia & the Pacific Policy Studies*, 2(3), pp. 1–14.

Hardin, G., 1968. The tragedy of the commons. *Science*, 162(3859), pp. 1243–1248.

Hardy, D., 1988. Historical geography and heritage studies, *Area*, 20(4), pp. 333–338.

Harper, D., n.d. a. Appear. In: *Online Etymology Dictionary*. Available at: www.etymonline.com/index.php?term=appearandallowed_in_frame=0 [Accessed May 2015].

Harper, D., n.d. b. Synecdoche. In: *Online Etymology Dictionary*. Available at: www.etymonline.com/index.php?term=appearandsynedocheandallowed_in_frame=0 [Accessed May 2015].

Harris, W., 1998. Creoleness: the crossroads of a civilization? In: K.M. Balutansky and M. Sourieau, eds, *Caribbean Creolization: Reflections on the Cultural Dynamics of Language, Literature, and Identity*. Barbados: Press of the University of the West Indies, pp. 23–36.

Harris, W., 1999. *Selected Essays of Wilson Harris Volume 1* (A.J. Bundy, editor). London and New York: Routledge.

Harrison, F.V., 1991. Women in Jamaica's urban informal economy. In: C. Mohanty, A. Russo and L. Torres, eds, *Third World Women and the Politics of Feminism*. Bloomington: Indiana University Press, pp. 173–196.

Harvey, D., 1993. From space to place and back again: reflections on the condition of postmodernity. In: J. Bird, B. Curtis, T. Putnam, G. Robertson and L. Tickner, eds, *Mapping the Futures: Local Cultures. Global Change*. London: Routledge, pp. 3–29.

Harvey, D., 1996. *Justice, Nature and the Geography of Difference*. Oxford: Blackwell.

Harvey, D., 2000. Cosmopolitanism and the banality of geographical evils. In: J. Comoraff and J. Comoraff, eds, *Millennial Capitalism and the Culture of Neoliberalism*. Durham, NC: Duke University Press, pp. 529–564.

Harvey, D., 2014. Heritage and scale: settings, boundaries and relations. *International Journal of Heritage Studies*, 21(6), pp. 577–593.

Hau'ofa, E., 1993. Our sea of islands. In: E. Hau'ofa, V. Naidu and E. Waddell, eds, *In a New Oceania: Rediscovering Our Sea of Island*. Suva: University of the South Pacific, pp. 148–161.

Hau'ofa, E., 1994. Our sea of islands. *The Contemporary Pacific*, 6(1), pp. 148–161.

Hau'ofa, E., 1998. The ocean in us. *The Contemporary Pacific*, 10(2), pp. 392–410.

Hay, J.E., Mimura, N., Campbell, J., Fifita, S., Koshy, K., McLean, R., Nakalevu, T., Nunn, P. and De Wet, N., 2003. *Climate Variability and Change and Sea-level Rise in the Pacific Islands Region: A Resource Book for Policy and Decision Makers, Educators and other Stakeholders*. Apia, Samoa: South Pacific Regional Environment Programme.

Hay, P., 2002. *Main Currents in Western Environmental Thought*. Sydney: University of New South Wales Press.

Hay, P., 2006. A phenomenology of islands. *Island Studies Journal*, 1(1), pp. 19–42.

Hay, P., 2013. What the sea portends: a reconsideration of contested island tropes. *Island Studies Journal*, 8(2), pp. 209–232.

Hayward, P., 2012. Aquapelagos and aquapelagic assemblages. *Shima: The International Journal of Research into Island Cultures*, 6(1), pp. 1–10.

Head, L., 2000. Renovating the landscape and packaging the penguin: culture and nature on Summerland Peninsula, Phillip Island, Victoria, Australia. *Australian Geographical Studies*, 38(1), pp. 36–53.

Hegarty, D. and Tryon, D., 2013. *Politics, Development and Security in Oceania*. Canberra: ANU E-Press.

Heidegger, M., 1951. Building, dwelling, thinking. In: D.F. Krell, ed., *Basic Writings: Martin Heidegger*. London and New York: Routledge.

Henke, H., 1997. Towards an ontology of Caribbean existence. *Social Epistemology: A Journal of Knowledge, Culture and Policy*, 11(1), pp. 39–58.

Hennessy, E. and McCleary, A., 2011. Nature's Eden? The production and effects of 'pristine' nature in the Galápagos Islands. *Island Studies Journal*, 6(2), pp. 131–156.

Hess, D., 2009. *Localist Movements in the Global Economy: Sustainability, Justice, and Urban Development in the US*. Cambridge, MA and London: MIT Press.

Hill, S. and Hill, J., 1987. *Richard Henry of Resolution Island*. Dunedin: John McIndoe.

Hogan, C., 2008. Kiribati Adaptation Programme (Stage II): Pilot Baseline Study Report: Survey of Public Awareness of and Attitudes towards Climate Change Issues and Challenges. Available at: www.climate.gov.ki/wp-content/uploads/2013/05/Survey-of-public-attitudes-to-climate-change-baseline-study.pdf [Accessed 11 September 2014].

Holdgate, M., 2013. *The Green Web: A Union for World Conservation*. London: Earthscan Publications.

Holm, W., 2002. *Eccentric Islands: Travels Real and Imaginary*. Minneapolis, MN: Milkweed Editions.

Hosein, G.J., 2012. Transnational spirituality, invented ethnicity and performances of citizenship in Trinidad. *Citizenship Studies*, 16(5/6), pp. 737–749.

Hsiung, P-C., 1996. *Living Rooms as Factories: Class, Gender, and the Satellite Factor System in Taiwan*. Philadelphia, PA: Temple Press.

Hubbard, J., 2013. Mediating the North Atlantic environment: fisheries biologists, technology, and marine spaces. *Environmental History*, 18(1), pp. 88–100.

Hubbard, P., 2008. Here, there, everywhere: the ubiquitous geographies of heteronormativity. *Geography Compass*, 2(3), pp. 640–658.

Hughes, T., 2011. Unfinished business: KAP II Implementation Completion Report for Government of Kiribati. Unpublished Internal Report, Office of the President, Tarawa, Kiribati.

Hunte, K., 2001. The struggle for political democracy: Charles Duncan O'Neal and the Democratic League. In: D. Marshall and G.D. Howe, eds, *The Empowering Impulse: the Nationalist Tradition of Barbados*. Kingston, Jamaica: Canoe Press, pp. 133–149.

Ibbotson, J., 2001. *Lighthouses of Australia: Images from the End of an Era.* Melbourne: Australian Lighthouse Traders.

Illich, I., 1973. *Tools of Conviviality.* Berkeley, CA: Heyday Books.

Index Mundi. 2014. Barbados Demographics Profile. Available at: www.indexmundi. com/barbados/demographics_profile.html [Accessed 28 December 2015].

Ingold, T., 2000. *The Perception of the Environment: Essays in Livelihood, Dwelling and Skill.* London and New York: Routledge.

Ingold, T., 2007. Earth, sky, wind and weather. *Journal of the Royal Anthropological Institute,* 13(Issue Supplement), pp. s19–s38.

Ingold, T., 2010. Footprints through the weather-world: walking, breathing, knowing. *Journal of the Royal Anthropological Institute,* 13(Issue Supplement), pp. s19–s38.

INSEE [Institut national de la statistique et des études économiques], 2005. Estimation du PIB de Saint-Barthélemy et de Saint-Martin. Les documents de travail de CEROM n°2. Available at: www.insee.fr/fr/themes/document.asp?ref_id=13195&reg_id=26 [Accessed 31 December 2014].

International Labour Organization (ILO), 2007. Available at: www.ilo.org/global/topics/cooperatives/lang--en/index.htm.

Irwin, G., Ladefoged, T. and Wallace, R., 1996. Archaeological fieldwork in the inner Hauraki Gulf 1987–1996. *Archaeology in New Zealand,* 39(4), pp. 254–263.

Isern, T., 2002. Companions, stowaways, imperialists, invaders: pests and weeds in New Zealand. In: E. Pawson and T. Brooking, eds, *Environmental Histories of New Zealand.* Auckland: Oxford University Press, pp. 233–245.

Isserles, R.G., 2003. Microcredit: the rhetoric of empowerment, the reality of development as usual. *Women's Studies Quarterly,* 31(3/4), pp. 38–57.

Jackson, L., Lopoukhine, N. and Hillyard, D., 1995. Ecological restoration: a definition and comments. *Restoration Ecology,* 3(2), pp. 71–75.

Jazeel, T., 2014. Subaltern geographies: geographical knowledge and postcolonial strategy. *Singapore Journal of Tropical Geography,* 35(1), pp. 88–103.

Jessop, B., 2001. Institutional re(turns) and the strategic-relational approach. *Environment and Planning A,* 33(7), pp. 1213–1235.

Johnson, L.C., 2008. Re-placing gender? Reflections on 15 years of *Gender, Place, and Culture. Gender, Place, and Culture: A Journal of Feminist Geography,* 15(6), pp. 561–574.

Johnson, N.C., 2004. Social memory. In: J.S. Duncan, N.C. Duncan and R. Schein, eds, *Companion to Cultural Geography.* Oxford: Blackwell Publishers, pp. 61–68.

Johnson, N.C., 2009. Heritage. In: D. Gregory, R.J, Johnston, G. Pratt, M.J. Watts and S. Whatmore, eds, *The Dictionary of Human Geography.* Oxford: Blackwell Publishers, pp. 327–328.

Johnston, D.M. and Saunders, P.M., 1987. *Ocean Boundary Making: Regional Issues and Developments.* London: Croom Helm.

Karides, M., 2005. Whose solution is it? Development ideology and the work of micro-entrepreneurs in Caribbean context. *International Journal of Sociology and Social Policy,* 25(1/2), pp. 30–62.

Karides, M., 2007. Macro-economics and micro-entrepreneurs: comparing two island nations' responses to neo-liberalism and its impact on women's lives. In: A. Cabezas, M. Waller and E. Reese, eds, *The Wages of Empire: Neo-liberal Policies, Repression, and Women's Poverty.* Boulder, CO: Paradigm Publishers, chapter 5.

Karides, M., 2010. Theorizing the rise of micro-enterprise development in Caribbean context. *Journal of World Systems Research,* 17(2), pp. 192–216.

Karides, M., 2012. Local utopia as unobtrusive resistance: the Greek village micro-economy. *Journal of World Systems Research*, 18(2), pp. 151–156.

Kelman, I., 2007. Sustainable livelihoods from natural heritage on islands. *Island Studies Journal*, 2(1), pp. 101–114.

Kimber, C., 1966. Dooryard gardens of Martinique. *Yearbook of the Association of Pacific Coast Geographers*, 28, pp. 97–118.

King, R., 2015. Journal of Island Studies: preliminary ideas from 1991 and comments from 2015. *Island Studies Journal*, 10(1), pp. 7–10.

Kirby, V., 1996. Landscape, heritage and identity: stories from New Zealand's West Coast. In: C. Hall and S. McArthur, eds, *Heritage Management in Australia and New Zealand: The Human Dimension*. Melbourne: Oxford University Press, pp. 119–129.

Kiwis Against Seabed Mining, n.d. About KASM. Available at: http://kasm.org.nz/inside-kasm/about/ [Accessed 2 January 2015].

KNSO, 2012. *Kiribati 2010 Census Volume 2: Analytical Report (Vol 2)*. New Caledonia: Kiribati National Statistics Office and the Secretariat of the Pacific Community.

Kothari, U., 2005. Authority and expertise: the professionalisation of international development and the ordering of dissent. *Antipode*, 37(3), pp. 425–446.

Kothari, U., 2006. An agenda for thinking about "race" in development. *Progress in Development Studies*, 6(1), pp. 9–23.

Lansdown, R., 2006. *Strangers in the South Seas: The Idea of the Pacific in Western Thought: An Anthology*. Honolulu: University of Hawaii Press.

Lasserre, G., 1961. *La Guadeloupe*. Bordeaux: Union Français d'Impression.

Lassithotaki, A. and Roubakou, A., 2014. Rural women cooperatives at [sic] Greece: a retrospective study. *Open Journal of Business and Management*, 2(2), pp. 127–137.

Lattas J., 2014. Queer sovereignty: the gay and lesbian kingdom of the Coral Sea Islands. *Shima: The International Journal of Research into Island Cultures*, 8(1), pp. 59–71.

Laville, J.B., Levesque, B. and Mendell, M., 2007. The social economy: diverse approaches and practices in Europe and Canada. In: A. Noya and E. Clarence, eds, *The Social Economy: Building Inclusive Economies*, Paris: OECD, pp. 155–185. [Originally published in Montreal as *Cahier de l'ARUC=ES*, Cahier No C-11-2006.]

Lavoie, Y., Fick, C. and Mayer, F-M., 1995. A particular study of slavery in the Caribbean island of Saint Barthélemy: 1648–1846. *Caribbean Studies*, 28(2), pp. 369–403.

Lazarus, N., 2011. *The Postcolonial Unconscious*. Cambridge: Cambridge University Press.

Lefebvre, H., 1991. *The Critique of Everyday Life*. London: Verso.

Leontidou, L., 1990. *The Mediterranean City in Transition: Social Change and Urban Development*. Cambridge: Cambridge University Press.

Lewin, A., 2012. WAPA's solar power project moving ahead. *Virgin Islands Daily News*, 25 May. Available at: http://virginislandsdailynews.com/news/wapa-s-solar-power-project-moving-ahead-1.1320356 [Accessed 31 December 2014].

Leyshon, A. and Lee, R., 2003. Introduction. In: A. Leyshon, R. Lee, and C.C. Williams, eds, *Alternative Economic Spaces*. Thousand Oaks, CA: SAGE Publications, pp. 1–26.

Lloyd, G., 2013. Asylum seekers and the rhetoric of compassion. In: 2013 *James Martineau Memorial Lecture*. Hobart: University of Tasmania.

Loftsdóttir, K. and Palsson, G., 2013. Black on white: Danish colonialism, Iceland and the Caribbean. In: M. Naum and J.M. Nordin, eds, *Scandinavian Colonialism and the Rise of Modernity: Small Time Agents in a Global Arena*. New York: Springer, pp. 37–52.

Loizos, P., 1994. A broken mirror: masculine sexuality in Greek ethnography. In: A. Cornwall and N. Lindisfarne, eds, *Dislocating Masculinity: Comparative Ethnography*. London and New York: Routledge, pp. 66–81.

Longfellow, H.W., 1849. The lighthouse. In: *The Complete Poems of Henry Wadsworth Longfellow*, loc. 3212. [eBook, Public Domain]: Amazon.

Loreto, P., 2009. *The Crowning of a Poet's Quest: Derek Walcott's Tiepolo's Hound*. Amsterdam and New York: Rodopi.

Lovelace, E., 2013. Reclaiming rebellion. *Wasafiri*, 28(2), pp. 69–73.

Lowenthal, D., 1985. *The Past is a Foreign Country*. Cambridge: Cambridge University Press.

Lowenthal, D., 1997. Empires and ecologies: reflections on environmental history. In: T. Griffiths and L. Robin, eds, *Ecology and Empire: Environmental History of Setter Societies*. Melbourne: Melbourne University Press, pp. 229–235.

Maddern, J.F., 2008. Spectres of migration and the ghosts of Ellis Island. *Cultural Geographies*, 15(3), pp. 359–381.

Maddrell, A., 2013. Living with the deceased: absence, presence and absence-presence. *Cultural Geographies*, 20(4), pp. 501–522.

Maher, J., 2013. *The Survival of People and Languages: Schooners, Goats and Cassava in St Barthélemy, French West Indies*. Leiden: Brill.

Malpas, J., 2013. Heidegger, Aalto, and the limits of design. Available at: http://jeffmalpas.com/wp-content/uploads/2013/02/Heidegger-Aalto-and-the-Limits-of-Design.pdf [Accessed May 2014].

Marín, C., 2004. *Towards 100% RES Supply: An Objective for the Islands*. International Scientific Council for Island Development, European Island OPET.

Maritime New Zealand, n.d. New Zealand's search and rescue region. Available at: www.maritimenz.govt.nz/Commercial/Shipping-safety/Search-and-rescue/SAR-region.asp [Accessed 21 January 2015].

Massey, D., 1991a. A global sense of space. *Marxism Today*, 35(6), pp. 24–29.

Massey, D., 1991b. Flexible sexism. *Environment and Planning D: Society and Space*, 9(1), pp. 31–57.

Massey, D., 1994. *Place, Space and Gender*. Minneapolis: University of Minnesota Press.

Massey, D., 2005. *For Space*. Thousand Oaks, CA: SAGE Publications.

Matsuda, M.K., 2007. 'This territory was not empty': Pacific possibilities. *Geographical Review*, 97(2), pp. 230–243.

McCall, G., 1994a. Nissology: the study of islands. *Journal of the Pacific Society*, 17(2/3), pp. 1–14.

McCall, G., 1994b. *Rapanui: Tradition and Survival on Easter Island*. Honolulu: University of Hawaii Press.

McClelland, D.C., 1967. *The Achieving Society*. New York: Free Press.

McDowell, L., 1999. *Gender, Identity, and Place: Understanding Feminist Geographies*. Minneapolis: University of Minnesota Press.

McKinnon, M., ed., 1997. *New Zealand Historical Atlas*. Auckland: David Bateman.

McLean, G., 2000. Where sheep may not safely graze: a brief history of New Zealand's heritage movement 1890–2000. In: A. Trapeznik, ed., *Common Ground? Heritage and Public Places in New Zealand*. Dunedin: University of Otago Press, pp. 25–44.

McMahon, E., 2003. The gilded cage: from utopia to monad in Australia's island imaginary. In: R. Edmond and V. Smith, eds, *Islands in History and Representation*. London and New York: Routledge, pp. 190–202.

McMahon, E., 2013. Reading the planetary archipelago of the Torres Strait. *Island Studies Journal*, 8(1), pp. 55–66.

McMillen, H., Ticktin, T., Friedlander, A., Jupiter, S.D., Thaman, R.R., Campbell, J., Veitayaki, J., Giambelluca, T., Nihmei, S., Rupeni, E. and Apis-Overhoff, L., 2014. Small islands, valuable insights: systems of customary resource use and resilience to climate change in the Pacific. *Ecology and Society*, 19(4), Art. 44.

McNamara, K.E. and Gibson, C., 2009. 'We do not want to leave our land': Pacific ambassadors at the United Nations resist the category of 'climate refugees'. *Geoforum*, 40(3), pp. 475–483.

McSaveney, E., 2009. Nearshore islands – island sanctuaries. Available at: www.TeAra. govt.nz/en/nearshore-islands/5 [Accessed 31 January 2015].

MELAD [Ministry of Environment, Lands and Agricultural Development, Kiribati], 1999. *Kiribati Initial Communication under the UNFCCC*. Tarawa, Kiribati: MELAD.

Meredith, S., 2003. Barbadian Tuk music: colonial development and post-independence recontextualization 1. *British Journal of Ethnomusicology*, 12(2), pp. 81–106.

Meredith, S., 2004. Barbadian Tuk Music – a fusion of musical cultures. In: A.J. Randall, ed., *Music, Power, and Politics*. New York and London: Routledge, pp. 157–172.

Meredith, S., 2015. *Tuk Music Tradition in Barbados*. Aldershot: Ashgate.

Merleau-Ponty, M., 1945. *Phenomenology of Perception*, trans. C. Smith. London and New York: Routledge.

Météo-France, 2013. *Tableau des Alertes Cycloniques*. Available at: www.meteo.fr/ temps/domtom/La_Reunion/charte/pics/Alertes/tableau-alerte-cyclonique.html [Accessed 31 December 2014].

Mies, M., 1986. *Patriarchy and Accumulation on a World Scale: Women in the International Division of Labor*. London and New York: Zed Books.

Miller, C., Craig, J. and Mitchell, N., 1994. Ark 2020: a conservation vision for Rangitoto and Motutapu Islands. *Journal of the Royal Society of New Zealand*, 24(1), pp. 65–90.

Miller, E., 2006. Other economies are possible! Organizing toward an economy of cooperation and solidarity. *Dollars and Sense: Real World Economics*, July/August. Available at: www.dollarsandsense.org/archives/2006/0706emiller.html [Accessed 28 December 2015].

Monk, J. and Hanson, S., 1982. On not excluding half of the human in human geography. *The Professional Geographer*, 34(1), pp. 11–23.

Mosse, D., 2001. 'People's knowledge', participation and patronage: operations and representations in rural development. In: B. Cooke and U. Kothari, eds, *Participation: The New Tyranny?* London and New York: Zed Books, pp. 385–393.

Motutapu Restoration Trust, no date. Motutapu Restoration Trust Home Page. Available at: www.motutapu.org.nz [Accessed 31 January 2015].

Moulaert, F. and Oana, A., 2005. The social economy: third sector and solidarity relations – a conceptual synthesis from history to present. *Urban Studies*, 42(11), pp. 2037–2053.

Mountz, A., 2014. Political geography II: islands and archipelagos. *Progress in Human Geography*, 39(5), pp. 636–646.

Müller, M.M., 2013. Postcolonial bureaucracies: power and public administration in 'most of the world'. *Postcolonial Studies*, 16(2), pp. 233–242.

Mullings, L. 1999. Images, ideology and women of color. In: M. Zinn and B. Dill, eds, *Women of Colour in U.S. Society*. Philadelphia, PA: Temple University Press, pp. 265–289.

Murray, G.R., 2009. Environmental implications of plastic debris in marine settings – entanglement, ingestion, smothering, hangers-on, hitch-hiking and alien invasions. *Philosophical Transactions of the Royal Society B: Biological Sciences*, 364(1526), pp. 2013–2025.

Murray, W.E. and Overton, J., 2011. The inverse sovereignty effect: aid, scale and neostructuralism in Oceania. *Asia Pacific Viewpoint*, 52(3), pp. 272–284.

Naipaul V.S. 2012. *The Mimic Men*. London: Picador Kindle Edition.

Nasioulas I., 2012. *Greek Social Economy Revisited: Voluntary, Civic and Cooperative Challenges in the 21st Century*. Frankfurt: Peter Lang Verlag.

Nathan, S., 2009. Conservation – a history: the need for conservation. Available at: www.TeAra.govt.nz/en/conservation-a-history/1 [Accessed 31 January 2015].

Navarro, T., 2010. Virgin capital: foreign investment and local stratification in the US Virgin Islands. Unpublished PhD thesis, Duke University, Durham, NC.

Neate, R., 2013. Seabed mining could earn Cook Islands 'tens of billions of dollars'. *The Guardian*, 5 August. Available at: www.theguardian.com/business/2013/aug/05/seabed-mining-cook-islands-billions [Accessed 17 December 2014].

Neemia-Mackenzie, U.F., 1995. Smallness, islandness and foreign policy behaviour: aspects of island microstates foreign policy behaviour with special reference to Cook Islands and Kiribati. Unpublished PhD thesis, University of Wollongong.

New Zealand Parliament, 2013a. Hansard (Debates) Crown Minerals Amendment Act 2013 Amendment Bill In Committee, Third Reading, vol. 690, p.10277. Available at: www.parliament.nz/en-nz/pb/debates/debates/50HansD2013051800000004/crown-minerals-amendment-act-2013-amendment-bill—-in-committee [Accessed 21 April 2015].

New Zealand Parliament, 2013b. Hansard (Debates) Third Readings, vol. 689, p. 9358. Available at: www.parliament.nz/en-nz/pb/debates/debates/50HansD20130416 00000016/third-readings [Accessed 21 April 2015].

Nightingale, T. and Dingwall, P., 2003. *Our Picturesque Heritage: 100 Years of Scenery Preservation in New Zealand*. Wellington: Department of Conservation.

Notton, G., Stoyanov, L., Ezzat, M., Lararov, V., Diaf, S. and Cristofari, C., 2011. Integration limit of renewable energy systems in small electrical grid. *Energy Procedia*, 6, pp. 651–665.

Noxolo, P. and Preziuso, M., 2012. Postcolonial imaginations: approaching a "fictionable" world through the novels of Maryse Condé and Wilson Harris. *Annals of the Association of American Geographers*, 103(1), pp. 163–179.

Nunn, P.D., 2007. Managing the present and the future of smaller islands. In: I. Douglas, R. Huggett and C. Perkins, eds, *Companion Encyclopedia of Geography: From the Local to the Global*. London and New York: Routledge, pp. 799–819.

Nunn, P.D., 2009. Responding to the challenges of climate change in the Pacific Islands: management and technological imperatives. *Climate Research*, 40(2/3), pp. 211–231.

Nunn, P.D., 2010. Bridging the gulf between science and society: imperatives for minimizing societal disruption from climate change in the Pacific. In: A. Sumi, K. Fukushi and Hiramatsu, A., eds, *Adaptation and Mitigation Strategies for Climate Change*. Dordrecht: Springer, pp. 233–248.

Nunn, P.D., 2012. Understanding and adapting to sea-level rise. In: F. Harris, ed., *Global Environmental Issues*. Chichester: Wiley, pp. 87–104.

Nunn, P.D., 2013. The end of the Pacific? Effects of sea level rise on Pacific island livelihoods. *Singapore Journal of Tropical Geography*, 34(2), pp. 143–171.

Nunn, P.D., Aalbersberg, W., Lata, S. and Gwillian, M., 2013. Beyond the core: community governance for climate-change adaptation in peripheral parts of Pacific island countries. *Regional Environmental Change*, 14(1), pp. 221–235.

Nurse, K., 2015. At the cusp of key areas of transdisciplinarity. *Island Studies Journal*, 10(1), pp. 13–14.

Ó Direáin, M., 1986. Homage to John Millington Synge. In: T. Kinsella, ed. and trans., *The New Oxford Book of Irish Verse*. Oxford: Oxford University Press. Available at: www.aranisland.info/wordpress/the-art-of-aran-writing/#.VoHnr_l95D8 [Accessed 28 December 2015].

O'Crohan (Ó Criomhthain), T., 1929. *An t-Oileánach. Baile Átha Cliath: Ó Fallamhain* [published as *The Islandman*, trans. R. Flower. London: Chatto and Windus, 1934].

Oberst, A. and McElroy, J., 2007. Contrasting socio-economic and demographic profiles of two, small island, economic species: MIRAB versus PROFIT/SITE. *Island Studies Journal*, 2(2), pp. 163–176.

OECD, 2005. *The Paris Declaration on Aid Effectiveness and the Accra Agenda for Action. Development*. Paris: OECD.

Olaniyan. T., 1999. Derek Walcott: liminal spaces/substantive histories. In: B. Edmondson, ed., *Caribbean Romances: The Politics of Regional Representation*. Charlottesville: University Press of Virginia, pp. 199–215.

Olofsson, M., Sahlin, J., Ekvall, T. and Sundberg, J. 2005. Driving forces for import of waste for energy recovery in Sweden. *Waste Management & Research*, 23(1), pp. 3–12.

Olwig, K., 1980. National parks, tourism, and local development: a West Indian case. *Human Organization*, 39(1), pp. 22–31.

Olwig, K., 2009. Nature. In: R. Kitchen and N. Thrift, eds, 2009. *International Encyclopaedia of Human Geography*. Oxford: Elsevier.

Oxford English Dictionary, 2015. Insularity. Available at: www.oxforddictionaries.com/definition/english/insularity [Accessed 19 May 2015].

Painter, J., 2010. Rethinking territory. *Antipode*, 42(5), pp. 1090–1118.

Palan, R., 2003. *The Offshore World: Sovereign Markets, Virtual Places, and Nomad Millionaires*. Ithaca, NY: Cornell University Press.

Palmié, S., 2013. Mixed blessings and sorrowful mysteries. *Current Anthropology*, 54(4), pp. 463–482.

Pandey, A., 2013. Exploration of deep seabed polymetallic sulphides: scientific rationale and regulations of the international seabed authority. *International Journal of Mining Science and Technology*, 23(3), pp. 457–462.

Papataxiarchis, E., 1991. Friends of the heart: male commensal solidarity, gender, and kinship in Aegean Greece. In: P. Loizos and E. Papataxiarchis, eds, *Contested Identities: Gender and Kinship in Modern Greece*. Princeton, NJ: Princeton University Press, pp. 156–179.

Park, G. and Potton, C., 1995. *Nga Uruora – The Groves of Life: Ecology and History in a New Zealand Landscape*. Wellington: Victoria University Press.

Paterson, M., 2009. Haptic geographies: ethnography, haptic knowledges and sensuous dispositions. *Progress in Human Geography*, 33(6), pp. 766–788.

Peart, R., 2004. *A Place to Stand: The Protection of New Zealand's Natural and Cultural Landscapes*. Auckland: Environmental Defence Society.

Peat, N., 2007. Last, loneliest. *Heritage New Zealand*, 104, pp. 28–29.

Pelling, M. and Uitto, J.I., 2001. Small island developing states: natural disaster vulnerability and global change. *Environmental Hazards*, 3(2), pp. 49–62.

Petropoulo, C., 1993. Some historical notes on the Greek cooperative movement. Available at: http://base.d-p-h.info/es/fiches/premierdph/fiche-premierdph-744.html [Accessed 15 January 2012].

Phillips, V., 1977. *Romance of Australian Lighthouses*. Melbourne: Rigby.

Potter, R.B. and Pugh, J., 2002. Planning without plans and the neo-liberal state: the case of St Lucia, West Indies. *Third World Planning Review*, 23(3), pp. 323–340.

Pugh, J., 2001. On communicative action and power: The National Commission on Sustainable Development of Barbados. *Caribbean Geography*, 12(1), pp. 1–11.

Pugh, J., 2005a. The disciplinary effects of communicative planning in Soufriere, St. Lucia: governmentality, hegemony and space-time-politics. *Transactions of the Institute of British Geographers*, 30(3), pp. 307–321.

Pugh, J., 2005b. Social transformation and participatory planning in St Lucia. *Area*, 37(4), pp. 384–392.

Pugh, J., 2009a. What are the consequences of the 'spatial turn' for how we understand politics today? A proposed research agenda. *Progress in Human Geography*, 33(5), pp. 579–586.

Pugh, J., ed., 2009b. *What is Radical Politics Today?* London: Palgrave Macmillan.

Pugh, J., 2010. The stakes of radical politics have changed: post-crisis, relevance and the State. *Globalizations*, 7(1/2), pp. 289–301.

Pugh, J., 2013a. Island movements: thinking with the archipelago. *Island Studies Journal*, 8(1), pp. 9–24.

Pugh, J., 2013b. Speaking without voice: participatory planning, acknowledgment, and latent subjectivity in Barbados. *Annals of the Association of American Geographers*, 103(5), pp. 1266–1281.

Pugh, J., 2014. Resilience, complexity and post-liberalism. *Area*, 46(3), pp. 313–319.

Pugh, J., 2016. The relational turn in island geographies: bringing together island, sea and ship relations and the case of the Landship. *Social and Cultural Geography*, Epub ahead of print. http://dx.doi.org/10.1080/14649365.2016.1147064.

Pugh, J. and Momsen, J.H., eds, 2006. *Environmental Planning in the Caribbean*. Aldershot: Ashgate.

Pugh, J. and Potter, R.B., 2000. Rolling back the state and physical development planning: the case of Barbados. *Singapore Journal of Tropical Geography*, 21(2), pp. 183–199.

Pugh, J. and Potter, R.B., 2001. The changing face of coastal zone management in Soufriere, St Lucia. *Geography*, 86(3), pp. 247–260.

Pugh, J. and Potter, R.B., eds, 2003. *Participatory Planning in the Caribbean: Lessons from Practice*. Aldershot: Ashgate.

Purfield, C., 2005. *Managing Revenue Volatility in a Small Island Economy: The Case of Kiribati*. Washington, DC: International Monetary Fund.

Puri, S., 1999. Canonized hybridities, resistant hybridities: Chutney Soca, Carnival and the politics of nationalism. In: B. Edmondson, ed., *Caribbean Romances: the Politics of Regional Representation*. Charlottesville: University Press of Virginia, pp. 1–12.

Ramirez-Llodra, E., Tyler, P.A., Baker, M.C., Bergstad, O.A., Clark, M.R., Escobar, E., 2011. Man and the last great wilderness: human impact on the deep sea. *PLoS ONE*, 6(8), e22588.

Read, A.D., Phillips, P. and Robinson, G., 1998. Landfill as a future waste management option in England: the view of landfill operators. *The Geographical Journal*, 164(1), pp. 55–66.

Resource Management Act 1991. Available at: www.legislation.govt.nz/act/public/1991/0069/latest/DLM230265.html [Accessed 28 December 2015].

Richardson, B.C., 1983. *Caribbean Migrants: Environment and Human Survival on St Kitts and Nevis*. Knoxville: University of Tennessee Press.

Robequain, C., 1949. Saint-Barthélemy, terre française. *Cahiers d'Outre-Mer*, 2(5), pp. 14–37.

Robinson, W., 2014. *Global Capitalism and the Crisis of Humanity*. Cambridge: Cambridge University Press.

Rose, G., 1993. *Feminism and Geography: The Limits of Geographical Knowledge*. Minneapolis: University of Minnesota Press.

Royle, S.A., 1999. From Dursey to Darrit-Uliga-Delap: an insular odyssey. Presidential Address to the Geographical Society of Ireland. *Irish Geography*, 32(1), pp. 1–8.

Royle, S.A., 2001. *A Geography of Islands: Small Island Insularity*. London and New York: Routledge.

Royle, S.A., 2012. Lessons from islands, or islands as miners' canaries? In: G. Baldacchino, ed., *Extreme Heritage Management: The Practices and Policies of Densely Populated Islands*. Oxford: Berghahn Books, pp. 246–259.

Royle, S.A., 2014. *Islands*. London: Reaktion Books.

Royle, S.A., 2015. Navigating a world of islands: a 767 island odyssey. In: G. Baldacchino, ed., *Archipelago Tourism: Policies and Practices*. Farnham: Ashgate, pp. 19–31.

Safa, H., 1995. *The Myth of the Male Breadwinner: Women and Industrialization in the Caribbean*. Boulder, CO: Westview Press.

Sauer, J.D., 1969. Oceanic islands and biogeographical theory: a review. *Geographical Review*, 59(4), pp. 582–593.

Savory, E., 2011. Utopia, dystopia, and Caribbean heterotopia: writing/reading the small island. *New Literatures Review*, 47/48, pp. 35–56.

Schumacher, E.F., 1973. *Small is Beautiful: Economics as if People Mattered*. London: Blond and Briggs.

Scott, J.C., 1999. *Seeing Like a State: How Certain Schemes to Improve the Human Condition Have Failed*. New Haven, CT: Yale University Press.

Seamon, D. and Mugerauer, R., 1985. *Dwelling, Place and Environment: Towards a Phenomenology of Person and the World*. Dordrecht: Marinus Nijhoff.

Searle, G., 2013. *First Order: Australia's Highway of Lighthouses*. Adelaide: Seaside Lights.

Seaside-Lights, 2013. First order: Australia's highway of lighthouses. Available at: www.seasidelights.com.au/firstorder/# [Accessed 14 April 2014].

Secretariat of the Convention on Biological Diversity, 2009. Connecting Biodiversity and Climate Change Mitigation and Adaptation: Key Messages from the Report of the Second Ad Hoc Technical Expert Group on Biodiversity and Climate Change. Montreal, Canada. Available at: www.cbd.int/doc/publications/ahteg-brochure-en.pdf [Accessed 19 May 2015].

Selvon, S., 2006. *The Lonely Londoners*. London: Penguin Books.

Sen, A., 1985. *Commodities and Capabilities*. Oxford: Elsevier Science.

Shima Editorial Board, 2007. Editorial introduction to Island Culture Studies. *Shima: The International Journal of Research into Island Cultures*, 1(1), pp. 1–5.

Shukman, D., 2013. Deep sea mining 'gold rush' moves closer. *BBC News*, 17 May. Available at: www.bbc.co.uk/news/science-environment-22546875 [Accessed 10 November 2014].

Sidaway, J.D., 2002. Postcolonial geographies: survey-explore-review. In: A. Blunt and C. McEwan, eds, *Postcolonial Geographies*. London: Continuum Press, pp. 11–29.

Smith, A., 1759/2009. *The Theory of Moral Sentiments*. Cheyenne, WY: Uplift Publications.

Smith, L., 2012. Editorial. *International Journal of Heritage Studies*, 18(6), pp. 533–540.

Spivak, G.C., 1988. Can the subaltern speak? In: C. Nelson and L. Grossberg, eds, *Marxism and the Interpretation of Culture*. Chicago, IL: University of Illinois Press, pp. 271–313.

Srinivasan, U.T., Watson, R. and Sumailab, U.R., 2012. Global fisheries losses at the exclusive economic zone level, 1950 to present. *Marine Policy*, 36(2), pp. 544–549.

Stanley, K.M., 1991. *Guiding Lights: Tasmania's Lighthouses and Lighthousemen*. Hobart: St. David's Park Publishing.

Star, P., 2014. Review: island reserves and mainland islands, including a review of ecosanctuaries. *Environment and Nature in New Zealand*, 19(2), available at: http:// environmentalhistory-au-nz.org/2014/11/review-island-reserves-and-mainland-islands-including-a-review-of-ecosanctuaries/ [Accessed 28 December 2015].

Star, P. and Lochhead, L., 2002. Children of the burnt bush: New Zealanders and the Indigenous remnant, 1880–1930. In: E. Pawson and T. Brooking, eds, *Environmental Histories of New Zealand*. Auckland: Oxford University Press, pp. 119–135.

Stegen, K.S., 2015. Heavy rare earths, permanent magnets, and renewable energies: an imminent crisis. *Energy Policy*, 79, pp. 1–8.

Steinberg, P., Tasch, J., Gerhardt, H., Keul, A., Nyman, E. and Shields, R., 2015. *Contesting the Arctic: Rethinking Politics in the Circumpolar North*. London: I.B. Tauris.

Steinberg, P.E., 1999. Lines of division, lines of connection: stewardship in the world ocean. *Geographical Review*, 89(2), pp. 254–264.

Steinberg, P.E., 2001. *The Social Construction of the Ocean*. Cambridge: Cambridge University Press.

Steinberg, P.E., 2005. Insularity, sovereignty and statehood: the representation of islands on portolan charts and the construction of the territorial state. *Geografiska Annaler: Series B, Human Geography*, 87(4), pp. 253–265.

Steinberg, P.E., 2013. Of other seas: metaphors and materialities in maritime regions. *Atlantic Studies*, 10(2), pp. 156–169.

Storey, D. and Hunter, S., 2010. Kiribati: an environmental 'perfect storm'. *Australian Geographer*, 41(2), pp. 167–181.

Stratford, E., ed., 1999. *Australian Cultural Geographies*. Melbourne: Oxford University Press.

Stratford, E., 2003. Flows and boundaries: small island discourses and the challenge of sustainability, community and local environments. *Local Environment*, 8(5), pp. 495–499.

Stratford, E., 2008. Islandness and struggles over development: a Tasmanian case study. *Political Geography*, 27(2), pp. 160–175.

Stratford, E., 2013. Guest editorial introduction. The idea of the archipelago: contemplating island relations. *Island Studies Journal*, 8(1), pp. 3–8.

Stratford, E., 2015. A critical analysis of the impact of *Island Studies Journal*: retrospect and prospect. *Island Studies Journal*, 10(2), pp. 139–162.

Stratford, E. and Langridge, C., 2012. Critical artistic interventions into the geopolitical spaces of islands. *Social & Cultural Geography*, 13(7), pp. 821–843.

Stratford, E. and Low, N., 2015. Young islanders, the meteorological imagination, and the art of geopolitical engagement. *Children's Geographies*, 13(2), pp. 164–180.

Stratford, E., Baldacchino, G., McMahon, E., Farbotko, C. and Harwood, A., 2011. Envisioning the archipelago. *Island Studies Journal*, 6(2), pp. 113–130.

Stratford, E., Farbotko, C. and Lazrus, H., 2013. Tuvalu, sovereignty and climate change: considering Fenua, the archipelago and emigration. *Island Studies Journal*, 8(1), pp. 67–83.

Strauss A.L. and Corbin, J., 1990. *Basics of Qualitative Research: Grounded Theory Procedures and Techniques*. Thousand Oaks, CA: SAGE Publications.

Suwa, J., 2007. The space of Shima. *Shima: The International Journal of Research into Island Cultures*, 1(1), pp. 6–14.

Swyngedouw, E., 2013. Into the sea: desalination as hydro-social fix in Spain. *Annals of the Association of American Geographers*, 103(2), pp. 261–270.

Tasmanian Government Department of Primary Industries, Water and Environment, Tasmania, 2008. Cape Bruny Lighthouse. In Parks and Wildlife Service Visitor Information [online]. Available at: www.parks.tas.gov.au/index.aspx?base=2539 [Accessed May 2014].

Tasmanian Government Department of Primary Industries, Water and Environment, 2012. South Bruny National Park. In Parks and Wildlife Service Visitor Information [online]. Available online at: www.parks.tas.gov.au/index.aspx?base=3773 [Accessed May 2014].

Teaiwa, T.K., Tarte, S., Maclellan, N. and Penjueli, M., 2002. *Turning the Tide: The Need for a Pacific Solution to Counter Conditionality*. Suva, Fiji: Greenpeace Australia Pacific.

The Guardian, 2015. Editorial. The Guardian view on geography: it's the must-have A-level. 13 August. Available at: www.theguardian.com/commentisfree/2015/aug/13/the-guardian-view-on-geography-its-the-must-have-a-level [Accessed 6 December 2015].

Thom, D., 1987. *Heritage: The Parks of the People*. Auckland: Lansdowne Press.

Titchen, S., 1996. Changing perceptions and recognition of the environment – from cultural and natural heritage to cultural landscapes. In: J. Finlayson and A. Jackson-Nakano, eds, *Heritage and Native Title: Anthropological and Legal Perspectives*. Canberra: Native Title Research Unit, pp. 40–52.

Towns, D. and Ballantine, W., 1993. Conservation and restoration of island ecosystems. *Tree*, 8(12), pp. 452–457.

Towns, D., Wright, E. and Stephens, T., 2009. Systematic measurement of effectiveness for conservation of biodiversity in New Zealand. In: B. Clarkson, P. Kurian, T. Nachowitz and H. Reddie, eds, *Conserv-Vision: A Celebration of 20 Years of Conservation by New Zealand's Department of Conservation*. Hamilton: Conference Proceedings, pp. 1–22.

Trans-Tasman Resources Limited, 2013. Application for a marine consent to undertake a discretionary activity. [pdf] Environmental Protection Authority. Available at: www.epa.govt.nz/Publications/FinalApplicationForm.pdf [Accessed 2 February 2014].

Treadwell, J., 1994. *The Rangitoto Bach Settlements and Community Hall: An Architectural and Historical Appraisal*. Onehunga: Treadwell Associates.

Treadwell, J. 2005. Islands of architecture in the Hauraki Gulf. Unpublished Master of Architecture thesis, University of Auckland.

Truman Proclamation, 1945. (no. 2667). [online] Available at: www.trumanlibrary.org/proclamations/index.php?pid=252andst=andst1.

Tsai, H-M., 2015. Candid reflections on the *Journal of Island Studies*. *Island Studies Journal*, 10(1), p. 14.

Tsai, H.-M. and Clark, E., 2003. Nature–society interactions on islands: introduction. *Geografiska Annaler: Series B, Human Geography*, 85(4), pp. 187–189.

Tuan, Y.F., 1979. Space and place: humanistic perspectives. In: S. Gale and G. Olsson, eds, *Philosophy in Geography*. Dordrecht: D. Reidel Publishing Company, pp. 387–427.

Tunbridge, J. and Ashworth, G., 1996. *Dissonant Heritage: The Management of the Past as a Resource in Conflict*. Chichester: John Wiley and Sons.

Turnbull, J., 2004. Explaining complexities of environmental management in developing countries: lessons from the Fiji Islands. *The Geographical Journal*, 170(1), pp. 64–77.

UNEP, 1994. Programme of Action for the Sustainable Development of Small Island Developing States. Available at: http://islands.unep.ch/dsidspoa.htm [Accessed 4 May 2015].

UNEP/GRID-Arendal (United Nations Environment Programme/Global Resource Information Database), 2011. Continental shelf – the last maritime zone. Available at: www.grida.no/publications/shelf-last-zone [Accessed 28 December 2015].

United Nations Convention on the Law of the Sea, 10 December 1982 (A/Conf.162/122). Available at: www.un.org/depts/los/conventionagreements/conventionoverviewconvention.htm [Accessed 28 April 2015].

United States Energy Information Administration, 2013. US Virgin Islands Territory Energy Profile. Available at: www.eia.gov/state/print.cfm?sid=VQ [Accessed 31 December 2014].

United States Environmental Protection Agency, 2001. Virgin Islands Signs EPA Order to Clean Up Anguilla Landfill; Concern for Virgin Islanders' Health and the Environment Prompts Action. Available at: http://1.usa.gov/Zmh11B [Accessed 31 December 2014].

United States Virgin Islands Bureau of Economic Research, 2013. Virgin Islands Annual Tourism Indicators. Available at: www.usviber.org/A13 [Accessed 28 December 2015].

Urry, J., 1990. *The Tourist Gaze: Leisure and Travel in Contemporary Societies*. Thousand Oaks, CA: SAGE Publications.

Van Trease, H., 1993. *Atoll Politics: The Republic of Kiribati*. Christchurch: Macmillan Brown Centre for Pacific Studies and Institute of Pacific Studies.

Vannini, P., 2011a. Mind the gap: the Tempo Rubato of dwelling in lineups. *Mobilities*, 6(2), pp. 273–299.

Vannini, P., 2011b. Performing elusive mobilities: ritualization, play, and the drama of scheduled departures. *Environment and Planning D: Society and Space*, 29(2), pp. 353–368.

Viaspace, Inc., 2013. Viaspace releases update of 7 MW AD project in St Croix. *Biomass Magazine*, 25 April. Available at: http://biomassmagazine.com/articles/8906/viaspace-releases-update-of-7-mw-ad-project-in-st-croix [Accessed 31 December 2014].

Virgin Islands Port Authority, 2013. Website of the Virgin Islands Port Authority. Available at: www.viport.com/ [Accessed 31 December 2014].

VLIZ, 2014. Maritime Boundaries Geodatabase, version 8. [online] Available at: www. marineregions.org/ [Accessed 18 May 2015].

Walcott, D., 1974. The Caribbean: culture or mimicry? *Journal of Interamerican Studies and World Affairs*, 16(1), pp. 3–13.

Walcott, D., 1986. *Derek Walcott: Collected Poems 1948–1984*. London and New York: Faber and Faber.

Walcott, D., 1998. *What the Twilight Says: Essays*. London and New York: Faber and Faber.

Walcott, D., 2010. In conversation with Christian Campbell. Hart House Theatre, Toronto, 23 November. Available at: www.youtube.com/watch?v=d_6mgbRSUzo [Accessed 28 December 2015].

Wallace, A.R., 1880. *Island Life*. London: Macmillan.

Waterton, E., 2005. Whose sense of place? Reconciling archaeological perspectives with community values: cultural landscapes in England. *International Journal of Heritage Studies*, 11(4), pp. 309–325.

Webber, S., 2013. Performative vulnerability: climate change adaptation policies and financing in Kiribati. *Environment and Planning A*, 45(11), pp. 2717–2733.

Webber, S., 2015. Mobile adaptation and sticky experiments: circulating best practices and lessons learned in climate change adaptation. *Geographical Research*, 53(1), pp. 26–38.

Wesley-Smith, T., 2013. China's rise in Oceania: issues and perspectives. *Pacific Affairs*, 86(1), pp. 351–372.

White, I., 2010. Tarawa Water Master Plan: 2010–2030. Available at: www.climate. gov.ki/wp-content/uploads/2013/03/Tarawa-Water-Master-Plan-2010.pdf [Accessed 14 April 2014].

White, I., Falkland, T., Perez, P., Dray, A., Metutera, T., Metai, E. and Overmars, M., 2007. Challenges in freshwater management in low coral atolls. *Journal of Cleaner Production*, 15(16), pp. 1522–1528.

Wilcox, M., 2007. *Natural History of Rangitoto Island, Hauraki Gulf, Auckland, New Zealand*. Auckland: Auckland Botanical Society.

Williams, A.J., 2010. Beyond the sovereign realm: the geopolitics and power relations in and of outer space. *Geopolitics*, 15(4), pp. 785–793.

Wilmshurst, J., Anderson, A., Higham T. and Worthy, T., 2008. Dating the late prehistoric dispersal of Polynesians to New Zealand using the commensal Pacific rat. *Proceedings of the National Academy of Sciences of the United States of America*, 105(22), pp. 7676–7680.

Wockner, G., 1997. Policy Conundrums in the National Parks: Nature, Culture and the Wolves of Isle Royale. Unpublished PhD thesis, University of Colorado, Boulder.

Woolnough, A., 1984. *Rangitoto: The Story of the Island and its People*. Auckland: Angela Woolnough.

World Bank, 2011. Implementation Completion and Results Report for Kiribati Adaptation Program Phase II – Pilot Implementation Phase (KAP II). East Asia and Pacific Region. Available at: www-wds.worldbank.org/external/default/ WDSContentServer/WDSP/IB/2012/03/16/000356161_20120316010427/Rendered/ PDF/ICR17510P089320C0disclosed030140120.pdf [Accessed 11 August 2014].

World Bank, 2013. Kiribati Adaptation Programme Phase II. Available at: http:// go.worldbank.org/FS6I1WMSD0 [Accessed 21 May 2015].

Wright, E.O. 2010. *Envisioning Real Utopias*. London: Verso.

Wrighton, N. and Overton, J., 2012. Coping with participation in small island states: the case of aid in Tuvalu. *Development in Practice*, 22(2), pp. 244–255.

Yoffe, S., 1994. An historical ethnography of the holiday communities on Rangitoto in the interwar period. Unpublished Master of Arts thesis, University of Auckland.

Young, D., 2004. *Our Islands, Our Selves: A History of Conservation in New Zealand.* Dunedin: University of Otago Press.

Zinovieff, S., 1991. Hunters and hunted: Kamaki and the ambiguities of predation in a Greek town. In: P. Loizos and E. Papataxiarchis, eds, *Contested Identities: Gender and Kinship in Modern Greece.* Princeton, NJ: Princeton University Press, pp. 203–220.

Index

aborigines; *see also* Māori (*iwi*); resilience: Ngati Toa and Ngati Wai peoples of New Zealand 103; Nuenonne of Bruny Island 36, 38
The Advocate News (Barbados) 141
Agia Paraskevi collective 90, 94
Agra cooperative 89
Alpine Energy waste-to-energy proposal 124–7, 129
Anemotia cooperative 91
Anguilla Landfill 122–7, 130, 131
aquapelago neologism 5, 167
archaeologists, conflict with NZ environmentalists and ecologists 110–12
Asomotos women's cooperative 90
Auckland Star, 'Settler' articles 108–9
Australia; *see also* Bruny Island, Tasmania; Cape Bruny lighthouse: benefits from phosphate reserves of Nauru and Kiribati 77n3; Northern Territory moratorium on mining in coastal waters 19; relations with Kiribati 17, 64
aviation hazards of St Croix landfill 122–7

Bade, David 7–8, 145, 147, 151, 153, 157, 158
Baldacchino, Godfrey 4, 42, 58, 132, 149, 160, 168
Barbados Landships 8, 133–4, 137–42
Bhabha, Homi 139–40
bird strikes as aviation hazard 122–4, 127, 131
Black Power movement 135–6
Blum, Hester 6
boundaries; *see also* Exclusive Economic Zones (EEZs): delineated by lighthouse 47, 51, 162; of islands as

aid to conservation 113; of islands real 156–7; provided by social constructs 32–3, 50; seabed mineral development and 10, 12–14, 17; trope of miniature, bounded world 164
Brathwaite, Kamau (E.K.) 132, 138
bridges, nature of 33, 34
Brown, Mark 18
Bruny Island, Tasmania 35–7; *see also* Cape Bruny lighthouse
Building Dwelling Thinking see Heidegger, Martin
built structures 6, 33, 34; *see also* Cape Bruny lighthouse
Burrowes, Marcia 141

capacity: capacity-building 60–1; defining 54–5, 60, 76; donor reluctance to fund labour for project implementation 73, 77n6; population penalty faced by island microstates 61–4, 66–8, 75, 114–15
Cape Bruny lighthouse: construction 36–7; impressions of visitors 6, 37–43; lighthouse keeper's wife's recollections 40, 43–4, 47–8; making meaning 44–9, 50; photographs 34*f*, 42*f*, 44*f*; place and built structures 6; use of phenomenological method 6, 34–5, 37, 50–3
capitalism: 'decodes and deterritorializes' 20; Mediterranean's ways of opposing 95
Caribbean; *see also* Barbados Landships; St Barthélemy; St Croix; Walcott, Derek: islands as testing sites for energy systems 8, 114–16; island studies in 154; map showing locations of St Croix and St Barthélemy 116*f*

Casey, Edward 32, 45, 50
causation, avoiding claims of 156
Chatham Rise Benthic Protection Area
  (BPA) 29
Chatham Rock Phosphate Ltd (CRP),
  mining proposal 24*f*, 29
Chauvin, Sébastien 117
China and Taiwan compete for
  diplomatic recognition in Pacific 77n3
citizenship, St Barthélemy vs St Croix
  128–9
Clayton, Daniel 82–3
climate change: adaptation 77n5; *see
  also* Kiribati; Kiribati Adaptation
  Project (KAP); impacts on water
  supply 64, 67, 70–2, 115; island nation
  vulnerability to 6–7, 39, 56, 146–7,
  152, 168
Collins, J. 122–3
colonialism/postcolonialism; *see also*
  terms of island culture: authority
  exercised through farce and imitation
  139–40; impact on island cultures 8,
  16, 127–31, 148
composting pros and cons 121–2
Connell, John, *Islands at Risk* 168
consumption, impact on waste 117–18
Convention on the Law of the Sea *see*
  Law of the Sea (UNCLOS)
Cook Islands 18
cooperatives *see* Greece
Cousin, Bruno 117
Craig, John 111
Cranwell, Lucy 108
creolisation and hybridity as shorthand
  for creative cultural adaptation 136
Cresswell, Tim 32
Crown Minerals Act 23–5, 28
cultural heritage *see* Cape Bruny
  lighthouse; Lesvos; New Zealand
culture; *see also* terms of island
  culture: challenges of cross-cultural
  communication 61, 125–7; I-Kiribati
  fears of losing 69–70, 72–3;
  St Barthélemy vs St Croix 127–31;
  suppression of Caribbean island
  138, 140–1

Darwin, Charles 164
Davidson, Janet 111
Dean, Annika 6–7, 146–7, 149, 152–3,
  154–5, 156, 157, 158
Deffontis, Alexandra 119, 120

desalination 115, 116, 119–20, 131
development; *see also* capacity; economics;
  international development agencies;
  seabed mineral development and island
  governance: continental misperceptions
  of island microstates 55, 56–64; donor
  benefits from Pacific resources 77n3;
  participatory 143, 146, 151, 154
donors; *see also* international
  development agencies: PEFA
  assessments 62; requirements 6–7,
  54–5, 60–1, 63–70, 75–6

ecology/ecosystems *see* environment
economics; *see also* development:
  contrast of island economic forms
  with hegemonic discourse of
  development 85, 93–6; convivial 78,
  80–4, 93–5; St Barthélemy vs St Croix
  127–30; uncertainty regarding impacts
  of seabed mining 27
Ecosystem Based Adaptation 65, 77n5
EEZs *see* Exclusive Economic Zones
  (EEZs)
Elden, Stuart 15
embodiment of felt experience and
  knowledge 33–5, 43, 45–6
Emerson, Ralph Waldo, similarities with
  Walcott 142
energy *see* renewable energy; waste
  management
English-Caribbean authors wrestling
  with inherited terms 132
environment; *see also* climate change;
  wildlife, native, protection in New
  Zealand: Chatham Rise Benthic
  Protection Area (BPA) 29; concerns
  on Caribbean islands 121, 124–5,
  125–6, 130; ecological climax concept
  101; ecological restorations of NZ
  islands 98, 104–7, 110, 112–13; and
  geography 162–3; potential impacts of
  seabed development 19–28
Environmental Protection Agency
  (EPA): New Zealand 23, 27, 29;
  United States 126, 127
Europe, expansion of cooperatives 85
European Union, funding of SPC Deep
  Sea Mining project 22
Europeans: British in Tasmania 35–6;
  compensation for property on Kapiti
  Island 103; settlement of New
  Zealand 7, 98, 100, 107, 109–10,

111, 113; settlement of St Croix 128; tropical climates portrayed as unsuitable for 58–9

Exclusive Economic Zones (EEZs): configuration of space 30–1; Gilbert, Phoenix and Line Islands (Kiribati) 56; jurisdictions of convenience 14–15; New Zealand 11–12, 22–9; Pacific island nations 17–21, 59

export business of Lesvos women's cooperatives 91–2

Extreme Heritage Management (Baldacchino) 168

Farbotko, Carol 152

Farrell, Winston, 'The House of Landship' 141–2

Federal Aviation Administration (FAA) 122–4, 127, 130

feminism, island; *see also* Lesvos: defining 78–9; feminisms as frames of analysis 80–4

Fergusson-Jacobs, Editha (Nancy) 138

field work, importance of 153

Fielding, Russell 8, 145, 148–9, 151, 153, 155–6, 158–9

finance *see* development; international development agencies

fisheries, potential impacts of seabed mining 29

Fletcher, Lisa 45, 166

Fleury, Christian 168

food and cuisine as main product of Greek women's cooperatives 79, 86–7, 90, 91–4

Friendly Societies (Barbados) 137–8, 139

Gardener, M. 110–111

Garvey, Marcus 138

geography *see* human geography; island studies

geopolitics of New Zealand's deep sea mining strategy 22–9

Gilbert, Phoenix and Line Islands *see* Kiribati

Gillis, John 52

Global Environment Facility (GEF) 64–5

Goldsmith, Mark 63, 152

governance *see* seabed mineral development and island governance

Greece; *see also* Lesvos: convivial economics 78–84, 93–5; cooperatives 85–6, 90–1, 96n1, 96n2; economic challenges 80, 86, 93–4

Griffin, Clifford E. 138, 139, 141

Grimble, Arthur 164

Haenen, Rémy de 117

Haque, Tobias 62

Harris, Wilson 137

Harvey, David 32, 83

Hau'ofa, Epeli 165; on mobility of islanders 147–8; on sea as source of regional connection and sustenance 152–3, 157–8; on sea of islands 59, 157, 167; on smallness 16, 58, 59; on 'world of islands' 132

Hawaii 148, 149, 154

Hay, Peter 5, 40, 48, 50

Hayward, Philip 5, 166

Heidegger, Martin 6, 33–4, 52–3n1, 151

Henke, Holger 138

Henry, Richard 102

Hill, Sir Arthur 108

'history as myth' vs 'history as time' 134–7, 140

'The House of Landship' (Farrell) 141–2

Hubbard, Jennifer 14

human geography; *see also* island studies; place; space/spatiality: exploring place, space, and environment 3, 161–3; theories of place as core concern in 32–5, 49–53, 145, 149–51, 161–2

hurricanes and incinerator operation 120

hybridity: Caribbean as hybrid space 134; trope of postcolonial island culture 133, 136, 140

Illich, Ivan, *Tools of Conviviality* 84

immigrants and refugees *see* migrants/migration

impasse, trope of postcolonial island culture 133, 136, 138, 141, 142, 143

incineration 115, 118–22, 127, 131

Ingold, Timothy (Tim) 46, 47

international development agencies 6–7, 54–5, 60–1, 75–6, 143; *see also* World Bank

inventiveness, trope of postcolonial island culture 133, 137, 140–1

island/ocean geographies: island constellations/ocean relations 16–17; submerged terrain/submarine materialities 17–20; topologies and contiguities 20–2

island studies; *see also* terms of island culture: approaches 150–5; attraction

to 1–4, 145–7; contributions of human geography to 157–9; desired or expected outcomes 147–8; early history 163–4; influencing other fields to engage with 155–7; island geographies, retrospect and prospect 160–8; transcript of conversations, introduction 8–9, 144
Island Studies Journal 3, 165–7
islands: as heuristic devices 151–2; as intensifiers 4, 164; islandness vs insularity 76–7n1; "islands are natural" idea 97–8, 106–7, 112–13; physicality 156–7, 163; as sites for testing energy systems 114–16
'Islands of the World' conference (Vancouver Island, 1986) 166

Japan, aid given where fishing access provided 77n3
The Journal of Marine and Island Cultures 167

Kapiti Island, NZ 98, 103
Karides, Marina 7, 146, 147–8, 150, 154, 155, 158
Karori Wildlife Sanctuary, NZ 98
King, Russell 147, 151, 166
Kiribati: defence assistance by Australia and New Zealand 17; issues with donor requirements 6–7, 54–5, 60–1, 63–70, 72–6; map 57f; place and people 55–6; Revenue Equalisation Reserve Fund (RERF) 58, 77n4; studying climate finance in 147, 152, 154–5, 156; water supply challenges 62, 64, 67, 70–2
Kiribati Adaptation Project (KAP): over-ambitious and over-complicated 64–70, 75–6; unintended consequences 70–5

labour: convict/indentured/slave 36–7, 41, 126, 128; division of 81; export as means of foreign exchange 58; reluctance of donors to fund 73, 77n6; small populations limit specialization of 58
Lamming, George 132
Law of the Sea (UNCLOS) 12–15, 59
Lesvos: description 79–80; impact of migrant crisis 95, 148; women's cooperatives 7, 83, 86–95, 154
lighthouses *see* Cape Bruny lighthouse
Little Barrier Island, NZ 98, 101, 102–3

Lloyd, Genevieve 49
Loftsdóttir, Kristen 128
Longfellow, Henry Wadsworth, poems on lighthouses 44–5
Loreto, Paola 142
Lovelace, Earl 140–1
Lowenthal, David, *The Past is a Foreign Country* 97

Mackey, Moana 25, 28
Maddern, Joseph (Jo) F. 45, 48–9
Maddrell, Avril 47, 48
Magras, Bruno 119–20
Māori (*iwi*): burning and clearing of land 100; island reserves ignored cultural heritage of 101, 113; islands seen as connected, not isolated 153; removal from Little Barrier Island 102–3; settlement on Motutapu Island 109–10, 111; tribal claims to coastal seafloor 17
Massey, Doreen 32, 135
McCall, Grant 146
Mesotopos cooperative 90–1
migrants/migration 61–2, 95, 148, 149, 168
mimicry/mockery, trope of postcolonial island culture 133, 134, 136, 137–41
mineral development *see* seabed mineral development and island governance
MIRAB syndrome 58, 63
mobility of islanders 59, 147–8, 150, 159
mockery *see* mimicry/mockery
Molyvos cooperative 86, 88, 91, 94
Motutapu Island, NZ 109–12
Murray, Thérèse 6, 145, 149–50, 152, 156–7, 158

national parks movement 100–1, 108
natural heritage *see* New Zealand
Nauru, phosphate reserves 77n3
New Zealand: co-finance of Kiribati's KAP II 64; deep sea mining and development 11–12, 22–9, 30–1, 153–4; defence assistance to Kiribati 17; Department of Conservation 105–6, 109, 110–11; island natural heritage 98–106; island reserves and cultural heritage 7–8, 101, 103–12; "islands are natural" idea 97–8, 106–7, 112–13; map 99f; search and rescue services to Pacific islands 17
Ngati Toa and Ngati Wai peoples of New Zealand 102–3

nissology 7, 59–61, 75–6, 79, 146, 165
non-compliance with waste management
　procedures 120–1
Noxolo, Patricia 143
Nuenonne of Bruny Island 36, 38

ocean; *see also* island/ocean geographies;
　seabed mineral development and
　island governance: binaries of land
　and water 146, 152; exceeds human
　scales 16; perceptions at Cape Bruny
　37–41, 45–7, 51; and physicality
　of place 156–7; sea as source of
　connection 59, 152–3, 157–8, 168;
　territorial orderings via ocean spaces
　12–14, 30–1
O'Crohan, Tomás 165
Ó Direáin, Máirtín 165
Olaniyan, Tejumola 140
oppression and invention intertwined 140

Pacific Commission (SPC) 21–2
Pacific island nations: interconnections
　17, 21–2, 29; land and territorial
　areas 18*f*; potential gains from
　seabed development 18–19; territorial
　contiguity 20–2
Pacific Small Island Developing States
　(SIDS) 18–19
Pākehā (non- Māori (*iwi*); *see* Europeans
Palsson, Gisli 128
Papataxiarchis, Evthimios 88
Parakoila, village of Anatolians 92–3
Paris Declaration on Aid Effectiveness 61
participation; *see also* development,
　participatory: in island studies
　154–5; trope of postcolonial island
　culture 133, 151; used to mask power
　imbalances 64, 76
The Past is a Foreign Country
　(Lowenthal) 97
PEFA *see* Public Expenditure and
　Financial Accountability (PEFA)
pests: exotic flora and fauna introduced
　in New Zealand 100; removal from
　NZ islands 105–6, 109–11
phenomenological method 6, 34–5, 37,
　50–3, 145
physicality of islands and place 46, 50,
　52, 156–7, 163
place; *see also* Cape Bruny lighthouse;
　human geography; physicality of
　islands and place; space/spatiality:

commitment to community and 91–3;
　feminism, and economic development
　78, 80–4; social construction of 32–3,
　34, 46, 50–1, 79
Polichnitos cooperative 91
postcolonialism *see* colonialism/
　postcolonialism
Preziuso, Marika 143
prisons, islands as 47–8, 145, 158
PROFIT economic models 77n2
protests against seabed development 23–5
Public Expenditure and Financial
　Accountability (PEFA) 62
Pugh, Jonathan (Jon) 8, 146, 148, 149,
　151, 152, 154, 159

Rangitoto Island, NZ 107–9
Raynal, Guillaume-Thomas 128
Recreation Reserves *see* Rangitoto
　Island, NZ
recycling 118, 121, 131
religion *see* culture
renewable energy 114, 119–20, 124, 127
RERF *see* Kiribati, Revenue
　Equalisation Reserve Fund (RERF)
resilience: attributed to islanders 40,
　43, 49, 58, 59, 153; to climate change
　7, 72–3; trope of postcolonial island
　culture 133
Resolution Island, NZ 98, 101–2
Richardson, James 102
Royle, Stephen 3, 4, 9

St Barthélemy, waste-to-energy
　applications 115–22, 127–31
St Croix, waste management issues 116,
　122–31
St Lucia *see* Walcott, Derek
Sammler, Katherine Genevieve (Kate)
　5–6, 146, 150–1, 153–4, 156, 157
Samoa, defence assistance by Australia
　and New Zealand 17
scale; *see also* Kiribati: geography's 162;
　importance in island infrastructure
　systems 114, 119, 151; large vs small
　islands 149–50, 156; small-scale local
　economies 81–5, 95
Schumacher, E.F., *Small is Beautiful* 84
sea *see* ocean
seabed mineral development and island
　governance: background 10–12;
　corporate vs public interests 5–6;
　geopolitics of New Zealand's deep

sea mining strategy 22–9; history of offshore mineral desires 12–15; island/ocean geographies 16–22; sources of information 153–4; territorial orderings via ocean spaces 30–1
seawalls, donor-funded construction 64–7, 70–1, 74
Secretary Island, NZ 98, 102
Shima 165–6
silence related to moments of artificiality in postcolonial condition 134
SITE economic models 77n2
skills *see* capacity
slavery *see* labour
Small is Beautiful (Schumacher) 84
Small Island Cultures Research Initiative 166
smallness 16, 58, 59, 155–6
Smith, Adam 49
social construction: of ocean space 12, 13, 22; of place 32–3, 34, 46, 50–1, 79
social justice 145–7, 150, 155, 158
South Pacific Studies 167
space/spatiality; *see also* Exclusive Economic Zones (EEZs); place: of Caribbean life 134–7; gendered space in Greece 88–90; islands as relational spaces 142; large and small scales 162, 164; ocean exceeds human scales 16; spatial turn 155, 159; subalterity of island place 82–3, 145–6; territorial orderings via ocean spaces 12–14, 30–1
SPC *see* Pacific Commission (SPC)
Spivak, Gayatri C. 134
Steinberg, Philip E. 13, 46–7
Stratford, Elaine: attraction to islands and human geography 1–4; review of ten years of *Island Studies Journal* 166; transcript of conversations, preamble 144
suspension, trope of postcolonial island culture 133, 137–8, 141, 142–3
Swyngedouw, Eric 115
synecdoche, Cape Bruny as a 51
Synge, John Millington 165

Tabai, Jeremia 54
Tasmania; *see also* Cape Bruny lighthouse: Bruny Island 35–7; history as a penal colony 47
Te Kiri, Rahui, and Tenetahi, Te Heru 102–3

terms of island culture; *see also* Walcott, Derek: articulating life in islanders' own terms 8, 134–8, 140–3; artificiality of terms and postcolonial subjectivity 132–4; understanding islands/islanders on own terms 52, 55, 59–61, 147–51, 158, 165, 167
Thomas, Algernon P.W. 101–2
Thomson, George M. 101–2
Tools of Conviviality (Illich) 84
tourism; *see also* New Zealand, island natural heritage: island economies dependent on 127, 128; as market for Lesvos cooperatives 84, 86, 93–4; St Barthélemy 115, 117, 121, 122, 129, 130–1; St Croix 123–4, 126, 129–30
Trans-Tasman Resources Limited (TTR) mining proposal 24f, 26–7, 29
tropes and themes of postcolonial island culture 133, 136–8, 140–3, 151
tropical climates portrayed as unsuitable for Europeans 58–9
Tuan, Yi-Fu 45, 48
'Tuk' music 139, 140
Tuvalu 63

uncertainty: vs inevitability of seabed mining 25–7; projected onto islands 152
UNCLOS *see* Law of the Sea (UNCLOS)
United Nations, Convention on the Law of the Sea *see* Law of the Sea (UNCLOS)
United States Virgin Islands *see* St Croix
Urry, John 164

Vanuatu, moratorium on mining in coastal waters 19
Virgin Islands *see* St Croix

Walcott, Derek 8, 132–7, 138, 140, 142–3, 162–3
Wallace, Alfred Russel 163
waste management: islands as testing sites for energy systems 8, 114–16; issues on St Croix 116, 122–31; waste-to-energy applications on St Barthélemy 115–22, 127–31
water supply challenges; *see also* desalination: on islands 115, 130–1; in Kiribati 62, 64, 67, 70–2; pollution by landfill on St Croix 126; on St Barthélemy 117, 119, 120

'wet ontologies' 152
What the Twilight Says (Walcott) 134,
    136–7, 142
wildlife, native, protection in New
    Zealand 97–8, 102–6, 109–13, 113n1
women; *see also* feminism, island;
    Lesvos: absent or peripheral in field

journals from Cape Bruny 41; creation
    of income-generating activities 81–2
Wood, Moses (or Moses 'Ward') 138
Woolnough, Angela 108, 109
World Bank, climate finance in Kiribati
    6–7, 64–70, 72, 73, 74–5
World Wildlife Fund for Nature 110

9 781138 339354